Time Travel and Our Parallel Worlds

Part 3
All New In-Depth Real Life Stories In the News

Richard Bullivant

All Rights Reserved. No part of this publication may be reproduced in any form or by any means, including scanning, photocopying, or otherwise without prior written permission of Richard Bullivant.
Copyright © 2016

Table of Contents

Introduction
RAF Bentwaters
Circles
The Skinwalker Ranch
Doorways
Fata Morgana
Conclusion

Introduction

Time travel and parallel universes. The two subjects are inextricably intertwined. Consider this recent headline in the British newspaper, *The Telegraph*:

Parallel Universe Proof Boosts Time Travel Hopes

One of the great things about the concept of parallel worlds is that it can make time travel possible. Why? Because it gets rid of one of the most daunting paradoxes involved with time travel: the so-called 'Grandfather Paradox'. This suggests that time travel (into the past) would be impossible because anyone who travels backwards in time could take actions that would change things in the future. The classic example is a person going back and killing his or her own grandfather. That should be impossible because if a person killed his or her own grandparent, then that person would have never existed in the first place to become a time traveller!

But if our universe is actually composed of millions and billions and trillions of parallel universes, a person could travel backwards in time and 'land' not in his own timeline and universe, but in a parallel world, and one which is so close to the original, that it might be all but indistinguishable. Then all the 'benefits' of actual time travel could be made possible. So even if our murderous time traveller kills his own grandfather in the past, it would be no problem because his grandfather would only be dead in another 'side world' and would have lived out his life in the time traveller's world.

That is just one way that time travel and parallel universe theory intersect. A more paranormal example might be the ever present subject of ghosts. Think about it:

When a person dies, they instantly become a part of the past. Yet, if that person's ghost is seen in the present, the ghost is now a part of both past and present. One interesting way to deal with this conundrum is to place ghosts in a parallel world. When a ghost is sighted, the observer is actually getting a glimpse into a parallel universe in the present time.

And it gets more complex from there. But in this book we care less about the hard science and mind-numbing mathematics and physics of time travel and parallel universes, and more about the fact that people seem to be encountering both of these phenomenon almost every single day!

Both subjects are not only interlinked with each other, but also bring into play a variety of other paranormal issues, including:

UFO activity

Ghost and poltergeist phenomenon

Cryptozoology (Bigfoot, shape-shifting Skinwalkers, the Loch Ness monster)

Ancient monuments that act like portals to other worlds

Unexplained sightings of otherworldly cities and landscapes

… And more. All of the above are phenomenon that simply refuse to go away. They keep happening, science keeps denying it all, and yet all of it begs for a logical and, yes, even scientific explanation.

The fact is that by combining time travel theory with the possibility that parallel universes are real, offers an excellent explanation for all of the above. No longer do millions of people have to be written off as mentally unbalanced, delusional or crazy. A workable model for time travel and a structure of parallel universes not only explains why so many otherwise sane and normal people keep having strange experiences, but places an array of 'paranormal phenomenon' into the realm of 'normal phenomenon'.

In this book we will look at real world cases reported in the major mainstream media, which strongly suggests that all kinds of people from all walks of life have experienced encounters with time travel, parallel universes, and often both at the same time.

From one of the most famous UFO cases of all time, the *British Roswell*, to a major sighting of floating cities over China, time and again (no pun intended) people seem to be confronted with phenomena that have no current explanation other than those that involve time travel and parallel universe phenomenon. By the way, this includes the crop circle phenomenon, sightings of Bigfoot, bizarre encounters with 'shape shifters' in the deserts of the American southwest, and freaky experiences associated with stone monuments, both natural structures and manmade.

Not so long ago anyone considering time travel and parallel universes access as something real required a person to be so open minded, he or she might actually be considered open minded to the point of being empty headed. But we think the stories you will read here show that the pendulum is swinging in favour of what so many witnesses have been reporting in good faith. They insist that what they have experienced is real. At the same time, the standard dismissals of the hard-bitten, material-science sceptics are starting to wear thinner every day.

There is an old saying:

Call a man a dog once, and you insult him. Call him a dog a thousand times and he may start barking.

Well, one or two encounters with time travel or an occasional slip into an alternate universe can easily be written off as an anomaly, a hoax or mental imbalance of the witnesses. But when these kinds of things keep happening again and again, and to ordinary people of sound mind and good reputation - or thousands of people at once - even the sceptics may have to *start barking*.

Of course, we welcome you to be as sceptical as you like of these stories, while also keeping an open mind. NASA physicist Tom Campbell always recommends 'open-minded scepticism' as the best approach to exploring subject matter on the fringe. But

Campbell also says that this is where the real breakthrough in science has always come from - the fringe. He also says that we have to learn to live gracefully with uncertainty.

All good advice to remember as you confront the amazing stories you are about to read in this book.

RAF Bentwaters

It has come to be known as *Britain's Roswell* - for better or for worse.

The bizarre and spectacular events which occurred in late December of 1980 at a joint U.S. and British airbase known as RAF Bentwaters have become one of the most remarkable and important case studies for UFO researchers.

These researchers are convinced that what happened at this nuclear-weapons armed military base near England's Rendelsham Forest is proof that extraterrestrials are visiting earth, that they are real and that they have a profound, albeit unknown, bizarre agenda.

However, what many people may be unaware of is that the Bentwaters case took a dramatic turnabout thirty years after the actual event. This development involved dramatic new information provided by one of the direct witnesses to the UFOs encountered in Rendlesham Forest.

The new information strongly suggests that the Bentwaters-Rendlesham incident was not just about UFOs and possible aliens but rather is all about time travel.

Nevertheless, before we get to the time travel element, let us sketch out the basics of the amazing incidents that occurred in England's Rendlesham Forest just after the Christmas holiday of 1980.

The Bentwaters case is called the *British Roswell* because UFO researchers say the events there rival the significance of what happened in Roswell, New Mexico, in 1947, when local newspapers shocked the world by reporting that U.S. Air Force officials had recovered the remains of a crashed UFO. The headline in the *Roswell Daily Record* blasted:

RAAF(Roswell Army Airfield) Captures flying Saucer on Ranch in Roswell Region

But the Roswell story would quickly change a couple of days later as air force bigwigs quickly back-tracked on their original comments to reporters, releasing a statement that the only thing recovered near Roswell was a crashed military weather balloon.

Of course, after such a statement, subsequent cries of a government cover-up and endless conspiracy theories have raged ever since.

Just like the U.K.'s Bentwaters incident, the Roswell story grew more complex and expanded over the years – at least that is what the sceptics like to point out. They say the story has been embellished, added to and sensationalized, by not only hoaxers but also innocent people, who are simply gullible and who have misread the evidence of what really happened.

Sceptics say stories of a crashed UFO in the desert near Roswell soon morphed from crashed debris into stories about recovered bodies of dead alien pilots, and even stories of aliens who survived the crash and who then subsequently became secret prisoners of the U.S. military.

However, what most people do not know is that the Roswell incident remained largely forgotten for decades after the initial events of 1947. Almost no one talked about it or wrote about it, even within and among the UFO community. It was not until some thirty years later that the Roswell 'can of worms' was re-opened by the famed UFO researcher and nuclear physicist Stanton Friedman.

After intense research that included interviewing dozens of witnesses and poring over reams of government documents, Friedman became convinced that a real UFO had crash-landed in the New Mexico desert, that the U.S. Air Force had recovered the wreckage, that dead (and possibly living) alien bodies had been received and that the U.S. Government was now covering up the story.

Friedman spread his findings about the Roswell incident by writing articles, which appeared in mainstream press and UFO literature magazines, and by giving hundreds of lectures all over the world. Thus, the Roswell incident became a major element of the overall UFO narrative as we know it today.

Friedman began his one-man 'truth campaign' which began in 1978. The first 'conspiracy' book, *The Roswell Incident*, was published in 1980 by authors Charles Berlitz and William Moore.

Interestingly, the story has continued to develop ever since.

Bentwaters: The British Roswell

The UFO incident at RAF Bentwaters took a similar path to Roswell. At first, it was a little known event. What really happened was experienced by a handful of military personnel stationed at the military base in December 1980. The Air Force installation is located about eighty miles northeast of London and ten miles east-northeast of Ipswich, which is near Woodbridge in Suffolk, England.

RAF Bentwaters was a major base for the British during World War II. The British Government leased the base to the U.S. Air Force during the years of the Cold War, starting in about 1951, and it became home for the U.S. 81st Fighter Wing. The Americans also operated a base at Woodbridge and, along with Bentwaters, the two were known as the Twin Bases.

Bentwaters is situated next to Rendlesham Forest, which is a 1,500-hectare area of lovely mixed woodlands in Suffolk, owned and managed by the British Forestry Commission.

The Bentwaters base was a sensitive and very high security site for a variety of reasons, not the least being the numerous underground silos loaded with nuclear-armed missiles pointed at the Soviet Union – although this is something the U.S. and the British have always denied. Today, everyone (including retired high-ranking officers who were stationed there at the time) acknowledges that nuclear weapons were installed at the base.

Strange Lights at Night

Just after 3.00am in the early hours of the morning of 27 December 1980, an American security patrol reported that they saw a series of unexplained lights that appeared to be descending into nearby Rendelsham Forest. At first the soldiers thought they were witnessing a conventional aircraft of some kind that was having problems and which was coming in for an emergency landing. They notified their superior officers, who ordered a squad of soldiers to go out into Rendlesham to investigate.

A patrol of three men drove a Jeep out to the edge of the forest. Two of the men then went on foot into the trees to investigate. Some distance into the woods, the men saw a glowing object that appeared to be descending towards the ground. As they got closer, they could clearly see that it was a physical object - a nuts-and-bolts craft- of some sort, but certainly no conventional aircraft, or anything they could recognize. It is important to bear in mind at this point that these men were extremely familiar with all manner of military and commercial aircraft.

They described the object as being about the size of a small car or sedan. At first glance they reported it to be metallic in appearance, being fitted with a series of coloured lights, including a pulsing red light on top. A bank of blue lights was on the underside.

As the patrolmen continued towards the object it began to manoeuver through the trees moving away from them. Initially the men said they could hear what they thought were animals from nearby farms being driven into a frenzy, presumably due to panic. It was later determined that none of the local farms kept livestock, and the panicked animals were believed to be deer and other animals of the forest. After observing the object for an indeterminate time, it then rose to the top of the trees and vanished ... or so it was first reported.

The three men on the patrol were Sergeant Jim Penniston, Airman First Class John Burroughs and Airman First Class Ed Cabansag. It was Penniston and Burroughs who walked into the woods while Cabansag stayed behind in the Jeep in which they had driven to the edge of the forest. From the Jeep Cabansag could maintain radio contact with the base.

It did not come out until later - years later - that much more had happened on that first night of encounter with strange lights in Rendlesham Forest. Burroughs and Penniston saw and experienced far more than just strange lights moving through the trees.

However, before we get to that, let us review what happened on the subsequent two nights, which, according to the initial reports as filed by military personnel as otherworldly events, continued with high peculiarity on the following day, 28 December 1980.

On this night, base Deputy Commander Colonel Halt was at dinner when Lieutenant Bruce Englund interrupted the evening and said, "Sir, our UFO is back."

Englund reported that a number of base personnel had again witnessed a variety of strange lights moving around in Rendlesham Forest, just as they had the previous night. The lights were so dramatic that the soldier at first thought that perhaps the British had celebrated New Year's Eve earlier than the Americans by letting off fireworks, but no, clearly something far odder was going on.

Colonel Halt decided this time to go out personally to investigate. He changed his clothes and took with him three men to see the strange lights for himself. He was accompanied by Lieutenant Englund, Sergeant Monroe Nevels and Sergeant Bobby Ball.

The men brought along a Geiger counter, which is an instrument used to measure the levels of radiation. Colonel Halt wanted to take measurements near the area where Penniston and Burroughs had seen the lights the night before, and where an impression of a triangular landing site could be seen on the forest floor. The markings were three round impressions in the shape of a triangle.

While out in the woods, Colonel Halt and his men spotted two strange lights in the sky, that at first were star-like and then later resolved as larger balls of light. The objects appeared to be intelligently controlled, performing a variety of manoeuvers across the night sky, and were also seen to 'shoot beams of light' down to the ground. In fact, these beams of light came close to the location of the four men, sometimes grazing close to their feet. At times, the men reported seeing as many as five objects. Halt would later say he saw structured craft that simply could not have been of conventional or known design.

The events were recorded as they happened by a tape recorder carried by Colonel Halt. At one point Halt says one of the objects seemed to be in some kind of distress and it appeared as if molten metal was falling from it. Halt can also be heard describing what he was witnessing as 'unreal'. It is certainly clear from the excited tone of Halt's voice that something amazing was happening.

A Third Night

After the spectacular events of the previous night, the second night of the encounter, the UFOs would return a third time to Rendlesham Forest – and this is probably the most spectacular, but also most controversial night of the three-day event.

Although there are dozens of witnesses involved comprising of military, local police and ordinary citizens, the most sensational story comes from a man who was a young private at the time, Airman First Class Larry Warren, who was just eighteen years old.

Warren had been assigned for duty at RAF Bentwaters three weeks earlier. He had asked to work at RAF Bentwaters because his father had served at the U.S./British base years before him. Warren had been trained to work as part of the base security team and on that third evening was doing guard duty at the gates when odd things began to happen.

Airman Warren did not know what was going on, but he was suddenly ordered off guard duty and told by his superiors to join a team that was heading out into Rendlesham Forest. What Warren and many other military personnel then began to witness was extraordinary.

The men could clearly see moving lights within the forest and as they entered the woods, they saw a variety of objects. One of them was a triangular craft with a red light on top and a row of lights along the bottom – it seemed to be the same object that Burroughs and Penniston had seen on the first night, and which Colonel Halt and his squad had encountered on the second night. They also saw a glowing red sphere which 'exploded' right in front of them, sending a shower of red lights or particles to the earth below. Warren said that nobody in his team was hurt in any way by this explosion. There was a variety of other lights in the sky as well.

After the red ball of light exploded, Warren said the unit transformed and now became an object similar to the one observed earlier – once again the triangular shaped UFO with a red light on top and rows of lights along the bottom. As the stunned men stood dumbfounded unable to comprehend what they were seeing, the truly bizarre events of the night were about to happen.

Four years later, Larry Warren told it this way to television journalists from CNN:

"At that point we saw people, or whatever you want to call them, coming out ... there were no doors on this ship, these things moved out ... there were three of them, they had light around them. They came out, and as they stood there, to make it more complicated, they actually floated. They were 'beings'. They made no threatening motions. They did not communicate with me at all. Our base commander was there ... Lt. Col. Gordon Williams."

Gordon Williams initially denied he had been at the site on the night of the action, and he called Warren a liar. However, documents uncovered years later revealed that Gordon indeed had been present at the site on the night in question, and so it seems Warren was telling the truth after all.

Warren went on to say that a senior officer arrived and began an attempt at communication with the 'alien' beings. No exchange appeared to take place. In fact, Warren described the human-to-alien confrontation as a stand-off. He said he then returned to base while things were still happening behind him, arriving back at around 4.00am. He said the UFO remained in the area at the time he left.

After the encounter, Warren said he was 'debriefed' and forced to sign a statement saying that he saw nothing more than lights in the sky. He was also given strict orders to never speak of the events to anyone, not even with his fellow airmen, including those who were also witnesses to the whole event. In effect, he was told to completely forget everything that he had seen and was clearly told that not doing so would result in severe consequences, from being dishonourably discharged from the military, to possibly even a prison sentence.

Despite all, Larry Warren began speaking of his experience after he completed his service and received an honourable discharge, although he was still under a legal obligation to keep silent. He began talking to reporters and film crews, and telling his story at UFO conferences. Some fellow military personnel have claimed that he made the whole thing up, while others supported his story. The military never took legal action against Warren, but officially called him a liar.

What is clear is that the U.S. military made strenuous efforts to disregard the story, play it down and cover it up. That the military was actively covering up the story was the conclusion of CNN journalists who aired a documentary of the Bentwaters incident in 1984. Numerous documents later released through the Freedom of Information Act all but confirm that an official cover-up was enacted, and maintained.

An Unfolding Story – Time Shifts and Time Travel

What we have provided above is merely a brief overview of the Bentwaters-Rendlesham Forest UFO event which, as we have said, is a story of enormous complexity, and one that has only grown more rich and variegated over the years as additional witnesses have come forward, adding details and further eyewitness accounts to the narrative. (For example, both mainstream journalists and UFO researchers later flushed out civilian witnesses who had also been in proximity to the Rendlesham Forest that night, and they all reported sighting bizarre lights in the sky. This included a local veterinarian and motorists passing by on a nearby highway.)

However, what concerns us most here is the story of Sergeant Jim Penniston, who was among the first group of three men who went into the forest on that first night of sightings of strange lights in the woods.

In the initial report, Penniston and his fellow aviator John Burroughs told of seeing only strange lights moving through the woods, sighting the triangular craft in the air, and then returning to base. It is likely that Penniston and Burroughs were also 'debriefed' and severely warned by military command to never speak of what they experienced that night in the forest.

As highly trained, loyal, experienced soldiers bound by their oath of service, they obeyed orders.

Both Penniston and Burroughs were left deeply affected by what they had encountered, and not in a positive way. Burroughs developed hearts problems and Penniston seemed to develop a severe case of PTSD, or Post-Traumatic Stress Disorder. He also experienced sleeping problems and eventually sought medical help and therapy, even though the military refused to provide medical coverage for his care, since it was the position of the Air Force that 'nothing of significance happened' on that night.

Penniston pressed forward seeking private psychological and medical help. In the course of his therapy he underwent hypnotic regression in the hope that he might recall more about the night of his encounter and uncover the reasons for his inability to sleep and other troubling and psychological problems that had plagued him since that event.

The story that came out of Penniston under hypnosis was sensational.

Direct Encounter with 'Alien' Craft

It seems that Penniston and his fellow aviator John Burroughs did a lot more than chase a few strange lights through the Rendlesham Forest that night.

Under hypnosis, Penniston recovered a more complete memory of that strange encounter on the first night in the forest. Penniston said that he and Burroughs had come upon the same triangular shaped UFO that had been seen by so many other base personnel on the following two nights. At one point, the craft landed on the forest floor directly in front of them.

As they approached the craft, both men mentioned feeling many odd effects. They experienced abnormal tingling sensations on their hair, skin and clothing. It was as if the air was electrically charged. They described what also felt like a time distortion and said that time felt as if it had slowed. They found it difficult to move and felt as if they were walking through syrup. It was a dreamlike sensation. Another chilling characteristic they noticed was that the forest seemed to have become devoid of all sound.

Eventually, the two men came within just a few meters of the landed craft.

Penniston decided to step forward and get as close to the object as he could. With remarkable bravery, he walked right up to the strange object as if it were an ordinary parked car on the street. He reached out and touched the glowing object with his hands, describing a black surface that felt like hardened glass. Penniston carried a notebook in which he took down details of everything that was happening, including making sketches of the craft.

In his notebook, Penniston described the object this way:

Triangular in shape. The top portion is producing mainly white light, which encompasses most of the upper section of the craft. A small amount of white light peers out the bottom. At the left side center is a bluish light, and on the other side, red. The lights seem to be molded as part of the exterior of the structure, smooth, slowly fading into the rest of the outside of the structure, gradually molding into the fabric of the craft.

Penniston was also stunned to see that there were strange markings on the outside of the craft - strange symbols that were slightly raised above the otherwise smooth, black glasslike surface of the object.

When Penniston put his hand on the etched symbols, he said they felt similar to sandpaper when compared to the rest of the smooth surface. As he touched them he said that everything became a brilliant bright white and that suddenly he could neither see nor hear. He became engulfed and alone inside a dazzling white light.

This effect occurred for an indeterminable amount of time. Eventually his sight returned and he found himself still standing next to the craft facing the pictorial glyphs. At this point, the craft started to turn a vivid bright white colour and the light was so strong that Penniston feared the strange object might explode. Survival instincts kicked in and he took a defensive position nearby as the craft became engulfed in light.

The craft didn't explode, but it lifted off the ground some four feet, moved away, and manoeuvered between the trees. It then ascended to tree top level and disappeared in the 'blink of an eye'.

The next day as Penniston was resting in his room in a troubled state after a sleep-disturbed night he began to experience the persistent effect of 'seeing' ones and zeros, like computer binary code, running through his mind's eye. The streams of symbols kept going through his thought processes in an endless feedback loop. Troubled by the churning images of ones and zeros, which he had apparently received from touching the glyphs in some kind of mental 'download', he felt compelled to write them down in a notebook. Jim recorded the ones and zeros in his notebook just as he saw them going through his mind.

He also made detailed sketches of the craft, including accurate renderings of the glyphs and symbols he had seen and touched. Then he put away his notebook and more or less forgot about it – until years later.

Remember, this was the year 1980 and computers were not as pervasive in our lives as they are today. It never occurred to Penniston that the ones and zeros he had copied verbatim from his mental download might be an actual binary code, let alone actually mean something! In fact, Penniston said in many subsequent radio and media interviews that he had always assumed the pages of ones and zeros were 'just a bunch of gibberish'.

Astonishingly, Penniston's notebook would remain stored away in a box of his belongings for the next thirty years! It was not until American journalist Linda Mouton Howe made contact with him in 2010 that Penniston happened to mention, almost in passing, that he had kept a notebook that contained detailed drawings and notes of his encounter in Rendlesham Forest in 1980.

Now, in our modern computer age, it seemed obvious that Penniston's sixteen pages of ones and zeros was a classic download of binary code. With great excitement, Howe suggested that a computer expert look at the pages so that a translation of the code might be achieved. Penniston agreed, although he still believed that what he had spewed out of his feverish memory all those years ago would be nothing more than nonsense. However, the code contained a tantalizing message. Among the first section of the code the following information was deciphered:

Exploration of Humanity.
Continuous for planetary advance.
Fourth coordinate continuous.

Further analysis determined that the fourth coordinate referred to the element of time.

Near the end of the message came the statements:
Origin 52 0942532 N (north latitude) 1 3131269 W (west longitude).
Origin Year 8100.

Yes, it seems that the information Penniston received that night was suggesting that the craft was in fact not an alien spacecraft from some other planet or galaxy, but a vessel from the future – far, far in the future from the year 8,100.

Again, further investigation of the code information seemed to suggest that the reason for this time travel expedition was one of desperation. It would seem that the future human race had developed a serious problem, a genetic problem that was threatening the existence of the entire human species.

The information suggested that the craft Penniston had touched was a probe that was on a mission to obtain DNA and genetic samples from the 20th and 21st Centuries, and that these samples were needed to serve as 'patches' in the biological make-up of a dying race.

Support from Secret Remote Viewing program

After Penniston released the binary code information, and its meaning and content had been decoded, it created not only a sensation but also a landslide of new sceptics and critics. The story seemed just too sensational, too fantastic!

Many questioned why this bombshell information had to wait thirty years to be revealed. Some even questioned whether Penniston's notebook had really been created at the scene back in 1980 by suggesting he created it later for his own reasons, perhaps for self-promotion. However, the notebook and ink were analyzed by forensics and it was determined that it had indeed been created in 1980.

The case continued to become infinitely more complex and convoluted. Claims and counterclaims flew back and forth. Debunkers had a field day pointing out what they thought were holes in the stories of all players involved, from Jim Penniston and John Burroughs, to Colonel Halt and former airman Larry Warren. Supporters of the legitimacy of the case fired back to debunk the debunkers – and the debate over Bentwaters-Rendlesham has raged ever since.

The Involvement of the American CIA

In the ongoing investigation of the case, researchers uncovered a once secret, but now declassified document, from an entirely different source - the CIA and the U.S. Military Defense program of psychic spying known today as the Remote Viewing program.

In brief: Remote viewers were a cadre of elite, secret military intelligence spies who were trained by scientists to develop their psychic abilities to a very high degree. The remote viewing program was launched in the early 1970s, and the first experiments were conducted at SRI, the Stanford Research Institute in Stanford, California.

The purpose of the remote viewing program was originally to spy on the Soviet Union during the Cold War. American spies in the CIA and military knew that the Soviets already had a psychic spying program, and were worried that it might be something that was very real and effective. In fact, they knew it was working because the Soviets were inexplicably getting their hands on some of the deepest 'black projects' of the U.S. military. Thus, it was determined that the U.S. needed its own team of psychic spies to counter the efforts of the Russians.

In addition to spying on the Soviets, the CIA and U.S. military intelligence agencies often tasked remote viewers to reach out with their minds to snoop into other targets. It was discovered from declassified documents that six years after the UFO events at Rendlesham Forest someone in the CIA had asked the remote viewing program to take a look at RAF Bentwaters to try to determine what had happened on those dates in late December 1980.

While sceptics and military officials have maintained that the soldiers on that night saw nothing more than stars and the winking lights of a nearby lighthouse, it was obvious that someone in the CIA and military defense was still deeply concerned about what had really happened at Bentwaters six years earlier.

The remote viewer who was tasked with probing the Rendlesham event was a top operative, Lyn (Leonard) Buchanan, and who is today considered one of the greatest and most gifted of the secret psychic spies.

It is important to note that Buchanan 'worked the target' totally blind. That is, in 1986, he had no idea what he had been tasked to remote view. Buchanan says that he had never heard of the Bentwaters case and, in any event, he was given nothing to go on except a random number. He was handed a piece of paper with six randomly generated digits and was simply asked to explain what he saw.

Amazingly, Buchanan began to describe the Rendlesham events in a lot of detail just as they had happened and as reported by Jim Penniston and John Burroughs. He described seeing the triangular craft, accurately described the smooth glasslike surface of the craft, and even described in some detail the strange glyphs imprinted on the side of the craft. Buchanan also noted that touching the glyphs would result in 'communication' between the UFO and whoever touched it.

But Buchanan was far from finished. He was asked to describe what might be inside the craft. It was his perception that the vessel was empty, except for an all-pervasive ambient red light, but also that the object was definitely under the control of 'beings' located at some other very far away location. In other words, the craft seemed to be a remote-controlled probe.

When pressed to describe the 'beings' who were controlling the craft, Buchanan said that he perceived frail, soft-skinned beings that he described as 'ETs' or extraterrestrials. He described these ETs as young, green, thin, frail and hot.

When asked where these ET's were located, Buchanan found his mind transported into the hallways of a massive underground structure. He perceived some of the ETs as being inside a 'curved domed, big, roomy interior filled with yellow light' and that the inside of this structure was 'hot, moist and surrounded by solid stuff'.

It was Buchanan's impression that the room with the interior yellow light was some kind of incubator that was filled with young ETs. He felt that these ETs were waiting for a time when a location could be found for them to leave their surroundings so that they could then begin to live their lives in an environment that would be more suited.

Buchanan was tasked to go even further and locate the source of who, what or where the so-called incubator originated from, and here he perceived a large circular construct of some kind that had a green and moist interior – and inside here were a group of ETs who were 'frustrated' and 'waiting'.

Finally, Buchanan said that the area outside the domed object with the green interior was on a barren planet that was cold, red and dusty and which seemed to be the planet Mars. It was Buchanan's impression that the place he visited was far into the future.

In short: Buchanan seemed to have nailed it! Once again, it is highly significant to note that Lyn Buchanan had absolutely no previous knowledge of what he had been tasked to remote view – yet he came up with a narrative that confirmed the original reports of airmen Jim Penniston and John Burroughs, doing so in a great deal of detail. The mathematical chances of an independent remote viewer 'working a blind target' six years after the event and coming up with the same set of facts are millions, if not billions to one.

Was it a case of True Time Travel?

If the above is all-true the fallout and implications of the Bentwaters-Rendlesham incident are beyond astounding. It suggests that not only are there aliens, or highly evolved future human beings living a desperate existence beneath the surface of cold, dusty Mars, but that they have developed the ability to travel through time.

Of course, the Bentwaters incident remains among the most fantastic and controversial stories in all of UFO history. Today, more than thirty-five years after those strange events, the debate rages on. Sceptics are not buying any of it. However, the alternative explanations offered by them are increasingly seen as absurd and even desperate.

Colonel Charles Halt to this day maintains that what he saw was not merely 'stars and a flashing lighthouse in the distance' – as sceptics insist – but an actual structured alien craft of some unknown origin. He is also convinced that what he saw was of extraterrestrial origin and he has never wavered from this claim.

Nor have the many other witnesses, from Jim Penniston and John Burroughs to Larry Warren and a cadre of civilian witnesses. Since the events of 1980, many other soldiers and airmen - dozens of them - who were stationed at Bentwaters during those extraordinary three nights in late 1980 have also come forward to support the 'genuine UFO' narrative.

Combine this with the fully independent assessment of one of the military's top remote viewers, and you have what many believe to be the most rock solid case of actual UFO-human contact in the history of ufology - a contact event that may also have involved a desperate search through time by a future humanity on the brink of extinction.

CIRCLES

In 1974, two world famous astronomers got together to attempt to send a coded message to see if they could establish contact with any aliens that might be in a particular faraway star system 25,000 lights years away.

Astronomers Carl Sagan and Frank Drake developed the message, part of which was a stylized DNA strand, a caricature of a human being and a symbol of our solar system. It was beamed out to the M13 Globular Star Cluster in the constellation Hercules. They did it to commemorate the remodeling and reopening of the gigantic Arecibo Radio Telescope in Puerto Rico, being the world's largest single dish radio telescope.

Because the message was sent using radio waves, which travel at the speed of light, the communication should not have arrived for 25,000 years. Nevertheless, many people believe that a 'return message' was received just twenty-seven years later in the form of a crop circle.

How could a message that should have taken 25,000 years to get to its destination be 'answered' in just twenty-seven years? (Note: Drake does not agree the message was replied to at all, but others think they have proof he is wrong. Read on.)

So what about the time difference? Was the message intercepted by another group of ETs before it got to its destination? Perhaps some form of time travel was involved, or at least time manipulation? Or maybe the crop circle return message was nothing more than a hoax?

We will get back to the Sagan-Drake-Arecibo 'return message' later in this chapter. What this case implies, however, is of central concern to this book: The subject of time travel.

It seems that the phenomenon of crop circles has landed right at the centre of the debate about whether time travel is possible. As you will see, what researchers have found hidden in the pattern of crop circles may be the best indication that time travel is not only possible, but already a 'standard mode of transportation' for some mysterious, non-human life form intelligences.

Decoding the Meaning of Crop Circles

By now just about everyone has heard about crop circles. Perhaps there is no one who does not have an opinion about them. It's safe to say that the vast majority of the general public, and nearly all of the scientific community, consider them nothing more than clever hoaxes created by 'artists' and others with too much time on their hands.

However, there is a small core of people who believe that not all crop circles are created by hoaxers and that many peculiar facets found within crop circles, not to mention their extraordinary designs, make it impossible that all of them could be created by earth-bound people. In other words, a percentage of them are 'real'. By real, we mean that they are the production of some manner of non-human intelligence, whether that be ETs, beings from other dimensions, intelligent earth spirits - or maybe even time travellers.

One man who is convinced that crop circles are real is one of the world's foremost biochemists, Dr Horace Drew. After growing up in Florida, Drew earned his Ph.D in Chemistry at Caltech in 1981, where he was the first person to make high-resolution x-ray structures of DNA. He then spent five years between 1982 and 1987 doing post-doctoral research on DNA and chromosomes at the MRC Laboratory of Molecular Biology in Cambridge. He finished his career as Principle Research Scientist at the CSIRO Division of Biomolecular Engineering in Sydney, Australia.

Drew retired in his adopted country of Australia in 2010 and became interested in the UFO phenomenon while working with UFO investigator Bill Chalker.

In 2002 Dr Drew had travelled to Cambridge, England, to meet with fellow scientists to discuss something called 'curved DNA' and on the same day of their gathering, a mysterious crop circle appeared very near where the scientists were meeting.

Amazingly, the crop circle was a magnificent representation of curved DNA. It is known as the Crooked Soley formation. It lasted just a few hours before farmers moved in to harvest the crop but, luckily, good photographs from the air were taken before it was erased.

The Crooked Soley circle was 300 feet across and depicted an extremely sophisticated representation of an unbroken double helix strand of DNA. The pattern contained 1,296 curved squares, 504 of which were squares where the crop was left standing. The other 792 squares were flattened. The effect was an amazing 3-D depiction of DNA strands curving around a central, star-like nucleus.

It was all too much to be a coincidence for Dr Drew. How could anyone, especially a gaggle of bored hoaxers, have known that Drew and his fellow scientists were meeting just a short distance away to discuss curved DNA molecules? Furthermore, the Crooked Soley crop circle contained mathematical facets that only a few people in the world were in a position to notice, comprehend and derive meaning from – and those people were present at the meeting of DNA scientists in Cambridge on that day.

Dr Drew was convinced that it just could not have been an accident, or the work of hoaxers. The Crooked Soley circle launched him into an intense investigation of crop circles, a study of their symbolism, and especially the obvious fact that many crop circles clearly contained mathematically encoded messages that had real meaning.

Proof Positive of Time Travel

Time and time again, Dr Drew has identified specific crop circles that contained messages which predicted key events that were to happen in the near future, sometimes just a few days after the appearance of the crop circle. Other crop formations foretold of events months or years in advance.

One was a crop circle prediction of the massive BP oil spill that occurred in the Gulf of Mexico in 2010. Three years before, two amazing crop circles appeared in the English countryside amongst fields of wheat.

The first appeared on 16 July 2007 in Hinton Downs. The crop formation appeared to be a two-pronged globular mass spreading out over a wide area of the field. Three days later on 19 July, nearby on Martinsell Hill, another crop circle emerged - this one a dead ringer for the well-recognized BP 'Green Sunburst' logo.

When the BP oil spill happened in 2010, images taken from space satellites of the massive oil slick were remarkably similar to the image laid out by the Hinton Downs crop circle pattern. Dr Drew contends that the chances of an image of an oil spill positioned very near a second image of the BP logo simply could not be a coincidence. He says that it is a clear example that someone or some form of non-human intelligence was showing us that it could reliably predict future events and, therefore, this form must surely be a time traveller.

One such 'hit' might be explained away as lucky chance, but Dr Drew has identified and reported dozens of cases where crop circle patterns clearly show and predict future events in a very unambiguous fashion, and in such a way that rules out utterly and completely any chance of a hoax.

Secrets of the Nobel Foundation Hacked by Crop Circles

Dr Drew, along with many others, are confounded by the ability of crop circle makers to predict with 100% accuracy what new scientific discovery will be awarded the Nobel Prize, before it is actually announced.

Bear in mind that the knowledge of which major achievement will be awarded the world's greatest prize, the Nobel, is among the most heavily guarded secrets in the world. A statement on the Nobel Foundation's website says:

The statutes of the Nobel Foundation restrict disclosure of information about the nominations, whether publicly or privately, for fifty years. The restriction concerns the nominees and nominators, as well as investigations and opinions related to the award of a prize.

Yet, time after time, crop circles have appeared just weeks or months before the official announcement by the Nobel Committee, with these circles seeming to predict exactly what the award will be given for.

Let us look at this example. On 25 June 2010, a crop circle depicting a perfect representation of a graphene electrode for a solar cell appeared on England's White Sheet Hill. In October of that same year, researchers Andre Geim and Kostya Novoselov were awarded a Nobel Prize for their development of the graphene solar cell.

Absolutely none dispute that the White Sheet Hill crop circle is a depiction of a graphene cell. However, what no one has a good answer for is how it showed up four months before the Nobel Foundation awarded its discovery the top prize for physics.

On 15 August 2011, an amazing crop circle appeared in Wiltshire, England, on a farm site called the Jubilee Plantation. The design of the crop circle was a brilliant work of art, and the actual execution of the pattern was a work of subtle genius. Those who observed it on the ground said the pattern of wheat flowed so gracefully and was so gently layered in a repeating pattern, it was reminiscent of water flowing in a stream - something that simply could not have been accomplished with the crude rope-and-board flattening methods of the hoaxers, which leaves tell-tale marks of 'stomping' damage on the grain.

The real kicker, however, is the pattern itself. The Jubilee circle has a complex design that represents an extremely precise two-dimensional model of the electron-diffraction pattern of the icosahedral Holmium-Magnesium-Zinc quasicrystal. The pattern is perfectly contained within a boundary that circumscribes a two-dimensional image of a higher dimensional quasicrystal framework.

While the Jubilee crop circle was a magnificent work of art in its own right, it anticipated by fifty-one days the next Nobel Prize for chemistry. On 5 October 2011, Dr Dan Shechtman was awarded the Nobel for the discovery of the Holmium-Magnesium-Zinc quasicrystal - the exact pattern of which is represented in the Jubilee crop circle pattern.

On 2 August 2009, a crop circle appeared in an English field that was an exact depiction of the CCD digital sensor chip used in video cameras. In October of that year, the Nobel was awarded jointly to Willard S. Boyle and George E. Smith for the invention of the CCD sensor.

Once again, it beggars the imagination to think of how these three examples of Nobel Prize predictions could have been carried out by conventional means, especially considering the Nobel Foundation's notorious reputation for extreme secrecy. For the Nobel crop circles to be hoaxes, it would have to involve some sort of bizarre conspiracy - or perhaps an inside leak from within the Nobel staff, and then a further connection with people interested in creating crop circle hoaxes in England!

Even if we were to write off the three correct predictions of the 2009, 2010 and 2011 Nobel awards, sceptics would still have to explain dozens of other time-defying predictions by crop circle makers. We have already mentioned the Gulf oil spill prediction, and one does not have to look far for many more precise crop circle prognostications of random future events.

For example, on 28 May 2015, the Chinese *Eastern Star* ferry launched on the Yangtze River for a pleasure cruise. Tragically, a major storm and perhaps structural problems with the ferry caused the ship to sink suddenly four days later on 1st June

2015. The terrible result was the death of almost four hundred people, all of whom drowned. It was among the worst maritime disasters for China in decades.

On 8 May, about three weeks before the sinking of the *Eastern Star*, a crop circle appeared in a field near Beijing, China. The circle was the design of a classic lotus flower, a blossom of great cultural significance in China where Buddhism is a major religion. The crop circle design also had many other strange and telling attributes. For one thing, it looked slightly lopsided and irregular, as if one petal had fallen off or had gone missing.

The central point of the lotus flower crop circle was positioned inside a hexagon, which created a special right-angle triangle, but to complete the hexagon, an observer from above had to take into account two small buildings completely outside the crop circle, although these buildings were positioned nearby. The triangle created between the crop patterns and the buildings is a known or special kind of triangle called a '5-12-13' referring to the relative lengths 5, 12 or 13 of its three sides.

Finally, inside and among the leaves of the lotus flower, an additional faint image of a Buddha figure can clearly be seen.

Dr Drew, in studying the Beijing crop circle, deduced a mathematical code incorporated into its structure. Detailing the complicated technical analysis provided by Dr Drew is beyond the scope of this book, but the result was clear-cut: Encoded within the design of this lotus flower formation was the date: 1 June 2015 - the date of the sinking of the *Eastern Star*.

In Chinese culture, a Buddha figure positioned with a lotus flower can symbolize mourning, or 'weeping for the people'. Additionally, one petal of the lotus flower was missing, another indication that 'something is wrong'. It would therefore seem clear that here again is another astounding example of a crop circle formation whose makers have some ability to transcend time, or at least have the ability to predict future events with uncanny accuracy.

Of course, sceptics charged that Dr Drew was simply playing with numerology. That is, they say that his methods of breaking the code of the lotus flower symbol was bogus. Sceptics say that he 'massaged the numbers' in such a way, and kept shifting and shaping his figures until they said what he wanted them to say. They feel that he just kept rearranging numbers until he got them to spell out 1 June 2015 … that the numbers could be arranged to say anything!

Unfortunately for the sceptics, there is one last aspect of the lotus flower crop circle that proves beyond reasonable doubt that the circle could not have been made by hoaxers, and also that Dr Drew was not simply making up or imagining numbers.

The proof comes by way of the overview from space as provided by *Google Maps*, which take pictures of the earth's surface from satellites in low-earth orbit. When the Beijing lotus flower crop formation is viewed from space, something additional and stunning can be seen plain as day: The formation is positioned right next to a gigantic representation of the *Eastern Star* – the ship that sunk and killed four hundred passengers!

The image of the *Eastern Star* is created not by directed artificial means, but by the shape of an urban area as seen from a bird's eye view. This is one case when only seeing is believing - and anyone can easily do so by searching out the image in question on the Internet. The undeniable fact is, when seen from space, the lotus star crop circle is positioned next to gigantic land formations - both naturally occurring and human made (buildings and streets) - that is a 'dead ringer' for a representation of the ferry itself!

In a manner that is clever almost beyond imagining, the crop circle makers leveraged images that could only be realized from the extreme altitude of low-earth orbit and incorporated them into their overall message, which was a precise prediction of a future event. Therefore, it seems that the crop circle makers are not only space travellers, but time travellers too.

The lotus flower formation of Beijing is just one example of crop circles that predict future random events, and which employ natural formations in the landscape of the earth. Again, seeing is believing. The reader is encouraged to seek images online which show crop circles positioned next to natural formations which clearly represent or reference what codes embedded in crop circles are talking about - messages about future events.

What the Government Knows

The story of time-travelling crop circle creators is not only endlessly complex, but also relentlessly weird and fascinating. For example, it seems clear that certain officials in the highest levels of world governments have long been aware of the true significance and reality of crop circles, but they have worked tirelessly to keep the general public in the dark about it all.

Why?

That is not an easy question to answer, but firstly, let us look at the evidence concerning high-level officials of world governments who know that crop circles are not the work of hoaxers, but rather some advanced form of non-human, higher intelligence.

A prime example of a government effort to cover up the significance of crop circles can be seen in the case of the Beijing lotus flower crop circle which foretold the tragic fate of the *Evening Star* ferry.

When crop circle researchers cracked the code of the Beijing lotus flower, they began to ask additional questions. In the course of seeking further information about the Beijing crop circle, a professor at Beijing University (a friend and collaborator of Dr Drew) called an administrator at a government institute who managed the land where the crop circle was created. He was greeted with a patently ridiculous tale from government bureaucrats.

The Chinese authorities said that they were aware of the lotus flower crop circle, but that fifty farmers using ropes and flattening boards as a way to celebrate and encourage agricultural production had created it. The Beijing professor then asked to speak with some of the farmers who had supposedly worked on creating the lotus flower, but none could be found. The professor then asked for other evidence, such as photographs of the great event, that is, pictures of the celebratory day when 'fifty farmers' went out to make a crop circle to 'celebrate agriculture'. No such photographs existed.

The professor then travelled to interview the local residents and farmers of the area where the crop circle appeared, but he could not find a single person who knew what he was talking about. None of the local people or farmers knew anything about a 'day of celebration' for agriculture or the creation of a crop circle in commemoration.

In short, the official government line that the lotus flower crop circle foretelling the fate of the *Eastern Star* ferry was actually manmade was a clumsy, even desperate attempt to cover up the true supernatural origin of the formation.

Again, one asks, "Why?"

What does the Chinese government know about crop circles, and what could be the possible reason for taking such great pains to keep the average man on the street from knowing the truth about them?

Other prominent crop circle researchers and journalists have also reported evidence that perhaps all world governments are aware that crop circles are not only real, but are expending a great deal of time and effort to 'control' the information crop circles represent, and to keep the general public confused and in the dark about it all. Consider this statement from American journalist and Emmy-winning filmmaker Linda Moulton Howe:

Time travel came up as the source of global crop formations in 1993. I was in Wiltshire, England, to explore the crop circle mystery, at a time when strange patterns in cereal crops had been reported in twenty-three different countries. There were TV news reports about crop patterns on ABC's 20/20 or CBS's 60 Minutes, with headlines in newspapers or magazines.

I met a man who told me that he had first-hand information that the CIA was trying to photograph by satellite every crop formation in chronological sequence, so that computers could look for mathematical patterns. Then he startled me, by explaining that the agency's theory is that we are dealing with time travellers. Crop formations have something to do with measuring the effectiveness of their efforts to change Earth's timeline. One reason why time travellers might interact with Earth would be to avert future disasters.

(Source: www.earthfiles.com)

As we saw in the previous chapter, to answer some of the toughest questions the CIA confronts, it frequently employs the use of one of its most powerful top secret resources - remote viewers - the elite team of psychic spies it originally created to spy on the Soviet Union during the height of the Cold War. In 1992, it tasked remote viewers to provide insight into the nature of crop circles. The remote viewers provided this summary:

*Crop circles are created by a collection of alien races who appear capable of time travel.
*Their field pictographs serve primarily as reference marks for event-line orientation, and not as messages to man.
*Such reference marks are produced by aliens moving through space and time, who are monitoring the course of events on Earth.
*At least twelve different species of extra-terrestrials use these signatures. For each mark, its date of creation, geographical location, and geometry are recorded in a central registry. (Source: remoteviewing.com)

British MI5 Is Deeply Involved

A number of prominent British crop circle investigators have also concluded that their own government is knee-deep in extraordinary and extreme efforts to sow confusion among the general public about the true nature of crop circles. Like the Chinese and the Americans, the British Secret Service is determined to keep the general public ignorant about what is truly going on with these mysterious messages laid down in fields of grain.

In a September 2015 newspaper article published in *The Express*, a Manchester based newspaper, long-time crop circle investigator David Clayton told reporters that the British MI5 has been paying hoaxers for decades to make fake crop circles for the sole purpose of propagating confusion and doubt among the general public about the reality of the formations.

Clayton is also a former RAF engineer. In *The Express* article, Clayton was reported as saying that MI5 were 'paying people to muddy the waters'.

In the 1990s Clayton, partnered with fellow researcher Robert Hulse, began to make an intensive study of crop circles. The two men spent more than twenty years investigating the formation inch by minute inch, concluding that some 95% of the circles were definitely fakes, but that a small percentage were undeniably genuine.

However, both men also became aware that the British government was determined to discredit the notion that any crop circle might be real (that is, not manmade) and had been hiring operatives to make as many fake circles as they could.

In a film made by British researcher and filmmaker Richard D. Hall, (www.richplanet.net) Hulse and Clayton tell of many sightings of RAF Apache attack helicopters over the appearance of new crop circles. Sometimes these military helicopters appeared to be chasing 'golden balls of light'. Such balls, or 'orbs' have been frequently sighted in areas where crop circles appear just a short time later; in fact sometimes just minutes later.

Hulse spoke to filmmaker Richard Hall:

"Without any doubt the balls of light are a real phenomenon, definitely. In fact, by these fields we have Golden Ball Hill, which has been called Golden Ball Hill probably for centuries, the reason being, golden balls have been seen there for so long. The military and the government have known about the real phenomenon for many, many years, and there is probably *a need to know basis* going on here as well."

Hulse and Clayton go on to say in Hall's film that a prominent website dedicated to showing how fake crop circles are made, CirclesMakers.com, is actually a product of British intelligence disinformation efforts. Hulse said:

"In our opinion ... the site is all part of a disinformation campaign. Originally we feel, possibly funded by government ... there was a link on the *CircleMakers* website which was a recruiting banner which took us straight to a government recruitment site for MI5. We do believe the genuine crop circles are extremely important and the government has known this for a long time, and that is why ... government teams were inaugurated - to throw sand in the eyes of researchers so that they wouldn't be able to see easily which were genuine and which weren't."

We will leave the subject of government mischief involved in the crop circle issue here because it isn't necessarily the scope of this book to explore what some might consider as being conspiracy theories associated with the phenomenon.

However, the fact that crop circles show an indelible link to issues of time travel, and all that this implies, might certainly be something that would create great interest and concern at the highest level of the security apparatus of governments around the world.

The Sagan-Drake Message

Now let us return to the subject we started this chapter on - the message sent to the Hercules Globular Cluster by astronomers Carl Sagan and Frank Drake. The message was sent in 1974, and twenty-seven years later, a crop circle appeared in a field near the Chilbolton radio telescope in Hampshire, England. The new crop formation was initially seen on 21 August 2001, and was probably created the night before.

The original message sent by Drake and Sagan was in the form of a series of ones and zeros, classic binary computer code. When interpreted, the code formed a rectangular shape that contained a number of interesting facts about our earth and humans.

The decoded Drake/Sagan message forms a simple pictogram showing the position of the earth as the third planet from the sun, a stick figure that represents a human being, an image of the radio telescope that sent it, and other sundry details – including a representation of the double-helix DNA molecule.

The first interesting aspect of the crop circle 'return message' is that it appeared in a field next to a radio telescope installation. Was this an acknowledgement by the ETs that they knew we had sent the first message via a radio telescope? That seems obvious.

At first glance, the return message seemed almost the same as the original message when decoded - a rectangular shape with a number of simple images. However, differences quickly became apparent. For one, the image of the human stick man was replaced with a shorter image of a humanoid, but this one with a large alien type head, similar to those aliens popularly known as the 'Greys'.

In the original message, Drake and Sagan inserted code to tell our potential 'alien friends' that the average height of a human being on earth was five feet nine inches. The 'Grey' alien in the return message was coded to indicate that it was about three feet tall, on average.

The Drake/Sagan message included the atomic symbols carbon, nitrogen, hydrogen, oxygen and phosphorous to indicate the basic building blocks of life on earth. The return message in the crop circle added an additional element – silicon.

This is significant because scientists have long speculated that although carbon is the most fundamental element of all known life forms, only one other element might serve as the most basic element for potential alien life, and that is silicon.

The DNA pattern had also been altered in a significant way. An additional DNA strand was added, and the number of nucleotides in the DNA strand was altered. This seemed a clear indication that the aliens were telling us about their basic DNA structure, just as we had told them about our own.

While the original message indicated we are on the third planet from the sun, the return message highlighted both the fourth and fifth planets in its solar system.

In place of the original radio telescope image, the alien return message contained a pictogram of an entirely different kind of communication method. What this device is or represents is not well understood but, whatever it is, it is the tool by which the crop circle was created, and from which the alien message was sent.

One of the most astounding aspects of the Chilbolton crop circle pictogram was that it was accompanied by a second formation in the same field - a mysterious image of a face. There is much speculation about who or what the face actually represents. Dr Drew thinks it looks like none other than Carl Sagan himself! Others, however, think it looks like the famous *Face on Mars*. Still others think it represents the face of a human/alien hybrid, and possibly the face of the alien who composed and sent the Chilbolton return message.

As we said, astronomer Frank Drake completely rejected the Chilbolton crop formation as a legitimate message. For one thing, he said, the DNA sequence in the return message was wrong, and he called it 'chemically impossible'.

To this, Dr Drew (who spent his entire career working with DNA) said that Drake didn't know what he was talking about. Dr Drew said:

"Astronomers and astrophysicists do what they do well, but they don't know a thing about DNA. The novel DNA structure shown at Chilbolton appears to be something known from origin-of-life experiments on Earth: namely a chemical variant of DNA called 2', 5' with only six nucleotides per single-stranded turn. One would have to be an expert DNA scientist to know that, and not a physicist or astronomer."

Drake also objected to the reality of the Chilbolton formation because he said the code for the supposed 'alien DNA' did not include silicon within the DNA pattern, but again, Dr Drew said this did not matter because it was simply not necessary for silicon to be included in the DNA structure.

Another problem Drake had with the Chilbolton formation was that it was received in just twenty-seven years. This meant that the alien race who had got the message could not be more than thirteen light years away (thirteen light years to send and receive and thirteen light years for the message to be returned). There is no known star system that number of light years from earth.

However, this assumes that a supremely advanced alien technology has not been able to crack the speed-of-light barrier as established by Einstein (that is, nothing in the universe can travel faster than light). On the other hand, Dr Drew said the speed of light barrier is irrelevant in any case because, again and again, crop circle makers have shown the ability to transcend time by showing us symbolic representation of what will happen in the future, and with great accuracy.

So once more, we can see that the subject of time travel remains a central element of the strange, overwhelming and endlessly complex subject of crop circles.

Incidentally, it was Dr Drew who also decoded the series of ones and zeros that we spoke of in chapter one - the code that was downloaded into the mind of U.S. Air Force Sergeant Jim Penniston when he touched a UFO in England's Rendlesham Forest in 1980. As you will recall, that message also indicated that the senders of the message were time travellers, originating some 6,100 years in the future.

Conclusion

No doubt the subject of crop circles is endlessly complicated and controversial. A constant back and forth battle rages between sceptics and believers as to the true nature of these bizarre formations, about 90% of which appear in southern England farm country near other ancient circular monoliths, such as Stonehenge and the massive stone circle of Avebury.

However, sceptics have been unable to find good arguments against the very obvious and incontrovertible evidence that dozens, if not hundreds, of crop circles are cleverly encoded with information that predicts future events with pinpoint accuracy.

For Dr Horace Drew and others, it is a closed case. Whoever or whatever is making crop circles has, at the very least, the ability to see into the future. Nevertheless, Dr

Drew goes much further, stating his belief that the makers of crop circles are just what they seem to be - time travellers.

The Skinwalker Ranch

If there is anywhere on the planet where it can be said that the best evidence exists for doorways into parallel worlds, a remote location known as the Skinwalker Ranch is almost certainly the number one contender for the title.

The property in question, once known as the Sherman Ranch, is located in the state of Utah in the American southwest. It occupies a space of four hundred and eighty acres just southeast of the small town of Ballard. The ranch borders the reservation of the Native American tribe known as the Ute, an ancient people that have lived in this harsh semi-desert locale for thousands of years.

It is from the Ute tradition and folklore that the name Skinwalker comes from, although the name itself may more properly be attributed to the Navaho, also a local tribe indigenous to this vast region. The Navaho version of the Skinwalker name is 'Yee Naaldlooshii' which when translated means, 'With it, he goes on all fours'.

Essentially, the Skinwalker is a kind of evil spirit or apparition, the exact nature of which is difficult to pin down. Sometimes called a Navaho Witch, the Skinwalker seems to be part monster, part spirit, part animal, often humanoid but, most of all, a 'shape shifter'. It can take on many forms and has a vast array of strange powers.

The Ute and the Navaho were bitter enemies for centuries. Both peoples were a culture famous for their fierce and warlike prowess. Both proved to be ferocious enemies during the coming of the white settlers to the Americas. The European invaders eventually overcame the Navaho and Ute, but only with extraordinary difficulty, and also only because the Europeans were helped along by superior numbers and modern weaponry. However, before the coming of the white man to the New World, the Ute and the Navaho fought endless battles against each other with the Ute tribe finally succeeding in driving the Navaho out of their territory.

To get some measure of revenge, the Ute believe that the Navaho placed a curse upon them. They believed that this curse was the coming of the Yee Naaldlooshii, the Navaho Witch or, what it is known as today, the Skinwalker.

Nevertheless, what does all this have to do with parallel worlds, and possibly time travel as well?

Read on.

Modern Times – Strange Events

While the local Native Americans have talked in hushed tones of the Skinwalker for centuries, it was not until about fifty years ago that this ancient legend would gain the attention of the general public. It all began with a series of stories telling of extremely bizarre events published in the *Utah Desert News*, the major newspaper of Salt Lake City.

More stories appeared a short time later in another major newspaper, the *Las Vegas Mercury*.

However, the stories were all over the place. That is, they did not focus on the singular phenomenon of the apparition of the Skinwalker. There was much more - a seemingly endless array of unexplainable sightings: glowing orbs, classic flying saucer UFOs, and cryptozoological beasts of many descriptions, from Bigfoot-like creatures to giant lizards, lizard-men, humanoid 'walkers'. There were also flying objects of such strange shape and form that they defy easy description, but were not just cigar-shaped or saucer-like UFOs.

And yes, among the incredible bizarre events that seem to point directly to parallel universe activity, as you will read later on, there have been numerous incidents in which witnesses have reported observing strange tunnels of energy opening up in midair on the Skinwalker ranch. Still more witnesses have told stories of stumbling into these inter-dimensional portholes, only to be thrust into some frightening landscape in some other alien world.

For the local members of the Ute tribe, what the modern media discovered was not new – or even news to them. Tribal elders, shamans and medicine men were well versed in the variety of otherworldly happenings on their ancient land. Just about every member of the Ute nation has had personal experience of such things, as well as what their elders have taught them. Perhaps it was only when the white man arrived and established a more permanent presence in the land that what had always been 'the norm' began to become national and international news.

The stories really gained traction in the early 1990s when a high profile and respected TV journalist, George Knapp of KLAS-TV, took an interest in the subject and brought it to a mainstream television news audience. KLAS is one of the primary local TV news stations in Las Vegas, and a CBS affiliate. CBS is one of the original 'Big Three' major networks in the United States, and so stories presented by them gain added credibility.

The Ranchers

The modern saga of the Skinwalker ranch began with the Tom Gorman family, even though the previous owners, the Shermans, were well versed in the eerie incidents that frequented this spooky location. It was simply that the previous owners just did not achieve the press that the new owners, the Gormans, attracted.

Tom Gorman purchased the land in about 1994. He was a resident of another state at the time, but it had always been his dream to be an independent and successful rancher. He came to this remote Utah location with high hopes as he planned to raise specialty breeds of cattle and provide a good life for himself and his family. However, as it transpired, he was completely unprepared for what lay in store for him.

On the very first day he arrived with his livestock, strange things began to happen, although to use the word 'strange' perhaps just doesn't do the incident justice.

Gorman and his family were releasing cows into a fenced-in area when suddenly they saw a large wolf approaching them. Coyotes are common to this area of Utah and while wolves are not unheard of, they are rare. However, it soon became apparent that this was no ordinary wolf as it was about twice the size of any normal, large adult wolf.

Wolves are generally shy of human beings. They are smart enough to know, due to vast painful experience, that getting too close to man can often result in them being trapped, shot and killed.

However, the giant wolf approaching the Gormans on that day was not shy at all. In fact, it seemed to the world to be more akin to a friendly dog, though definitely a wolf in form. Gorman said the huge canine had an intelligent look on its face, as if it were open and curious about his new human neighbours. At one point, Gorman said someone even reached out to pet the beast on the head!

Unfortunately, trouble soon started. The gigantic wolf's next move was to saunter over to the fence, reach through the slats and grab a calf with its massive, toothy maw. That was it! Gorman ran to his truck, retrieved a high-calibre rifle and shot the wolf at close range, scoring a direct hit.

What was immediately bizarre about the situation, however, is that the bullet seemed to have no effect on the animal. After being shot almost at point blank range, the wolf merely gave Tom Gorman a quizzical look and loped off back the way it had come from. The Gormans followed, attempting to track the beast, only to find that its trail vanished in mid-step, as if it had simply disappeared or had simply flown away!

There was also no blood. Again, Tom Gorman shot the wolf at very close range with a powerful hunting rifle, but there was no blood at the point of contact, and no blood leading away from the site of the shooting.

As strange as this event was, Gorman would soon wish that all of the events he and his family were about to experience were as tame as that one. The following years saw an unending series of hauntings, animal mutilations, disappearing livestock, poltergeist activity inside their home, strange lights 'walking' across the rugged landscape, UFO sightings, floating orb activity – and even the appearance of dead people, or the voices of dead people floating overhead. Even those deceased people that they had known personally from years past seemed to find their way to the Gorman ranch house to frighten them at night. They also sometimes heard voices floating overhead, but speaking in strange languages. Even though they could not understand the words, it was clear the disembodied voices were laughing at them and mocking them.

A Big Name Takes Notice

It was stories like these in newspapers and on television that attracted the attention of an American billionaire by the name of Robert Bigelow.

Bigelow grew up in Las Vegas, Nevada. He had been fascinated with the idea of space travel since he was a young boy. Like many youths of the era, Bigelow was excited by the advent of the space age and the enormous effort by the United States and Soviet Union to land a man on the moon. Bigelow grew up determined to find his place in the exhilarating world of space exploration but, in order to take part, he firstly decided to concentrate on making as much money as he could so that he would be able to 'make things happen' and further his dreams of aerospace adventure. With this in mind, he plunged himself into the real estate business, and eventually established a major hotel chain in America, which eventually made him a billionaire.

Bigelow began using his vast financial resources to pursue his childhood dream, establishing Bigelow Aerospace in 1999. Among his major accomplishments was the development of an expandable modular space habitat called *The Genesis*. Originally designed and then abandoned by NASA, Bigelow bought the patent rights for an expandable space habitat concept. He has since successfully placed two of the structures in earth orbit, where they remain today.

It is probably not surprising that a dreamer (but also a doer) like Bob Bigelow became captivated by the concept of UFOs, and the possibility that our earth was being visited by extraterrestrial beings from other solar systems, or even other galaxies.

Because his aerospace corporation is in his native Nevada, it was probably inevitable that Bigelow would take a great interest when the local Las Vegas TV station and local newspapers ran stories about the strange activities not far away in the neighbouring state of Utah, and on the Skinwalker Ranch.

The reports coming out of the area were so intriguing and so credible, Bigelow opted to visit the ranch himself – and then made an offer to buy it. By the time Bigelow showed up at the doorstep of the Gorman ranch house, this simple, hard-working family were absolutely fed up. Indeed, they were at their wits end. They were exhausted from years of being frightened out of their minds by a steady procession of strange events - unimaginably bizarre encounters which tormented them on an almost daily basis.

At the time Bigelow made his offer to buy the ranch, the Gorman family had taken to sleeping together on the centre of their living room floor. Husband, wife and children did not dare sleep in their respective rooms what with all the happenings - the ghosts, apparitions, poltergeist activity, monsters showing up outside to peer through the windows, UFOs hovering overhead – it was all too much for them.

Tom Gorman was eager to sell out to the wealthy businessman, recoup his unfortunate investment in an obviously haunted ranch and get on with a normal life. However, after the sale to Bigelow, Tom Gorman agreed to stay on as a ranch manager, although he relocated his family to a new home more than twenty miles away.

Major Scientific Study Launched

Robert Bigelow intended to find the underlying cause of what was happening on the Skinwalker Ranch, and he was determined to do it by the book, that is, by applying the sound principles of mainstream science, together with an objective approach.

To this end, he established the National Institute for Discovery Science (NIDSci), an organization with a focus on studying 'fringe phenomenon' or paranormal phenomenon, especially matters concerning the UFO issue. It must be said though that the NIDSci was not just another amateur operation, such as MUFON, or one of the many paranormal investigation 'clubs' that have become popular in recent year. NIDSci was staffed by mainstream researchers with doctorate level university training with world-class reputations.

Bigelow had money for the best.

The scientists set to work wiring all areas of the ranch with sophisticated electronic monitoring equipment. They set up enough automatic cameras to blanket every location using sensors of all kinds, from night vision cameras to electromagnetic monitoring devices. They furnished a trailer bristling with computers, infrared sensors, geomagnetic anomaly readers, and a plethora of other gear. They maintained a twenty-four hour staffed presence on the property.

The scientists conducted a geomagnetic survey to see if they could discover some kind of natural gravitational anomaly. They surveyed and sampled every plant species on the ranch to check for the presence of naturally hallucinogenic substances, such as 'magic mushrooms' or other compounds that might somehow be causing people to have strange visions. They theorized that substances like this may have leached into the ground water but all the water tests came back normal, and no intoxicating plants were found. They interviewed hundreds of people in the region, including neighbours and anyone who had ever visited the ranch.

After an extremely thorough assessment of the situation, KLAS-TV reporter George Knapp said that it wasn't long before the scientists starting seeing the bizarre phenomenon themselves, including lights, orbs, monsters, UFOs and countless other unexplainable events. Knapp told this to a MUFON audience in a 17 June 2008 lecture:

"There was a Ph.D. physicist who had worked on classified programs for several military entities and intelligence agencies. He was out there on the property with Dr Colm Kelleher and they see this black thing in the trees, kind of an amorphous cloud. It begins circling them in a menacing way. The dogs that are with them are terrified. And this guy, this physicist, begins speaking in another voice. This thing had gone into his head, and it is telling Colm that they are watching him and that he is not welcome there. This physicist doesn't remember that happening, but for days later this thing sort of hung with him and terrorized him at other locations on the property."

Knapp then spoke of an account that sounded like it could only be classed as a parallel universe 'portal' incident:

"These are world class scientists we're talking about ... there was another incident where there were two guys up on a ridge with instruments and telescopes and two other guys who were down below. The two guys down below start to see this dirty snowball of yellow light hanging maybe a foot off the ground in the middle pasture ... it's getting bigger and bigger and they're talking on the walkie-talkies to the guys up on the ridge and say, 'Can you see this thing?' and they reply, 'Yeah we're watching it.' And this ball starts getting bigger, then it gets elongated until it looks like a tunnel ... and then the guys with the infrared said, 'Hey there's something inside this thing.'

"And, sure enough, this large humanoid creature starts wriggling through the tunnel from one side trying to get to the other, as if it is coming from somewhere else. It gets to the end of this tunnel of light. It stands up. It's like eight feet tall, black, featureless, and it jumps out. The tunnel collapses on itself and goes away, and this thing goes running up Skinwalker Ridge ... and the guys that are up there just about had to change their underwear because they're thinking it is coming up after them. But it didn't ... it just went poof!"

One notable member of the NIDSci team was retired U.S. Army Colonel John Alexander who earned a Ph.D. in sociology and education, and later became a Commander within the U.S. Military's elite Army Special Forces. His level of credibility was high; that is, here was a man who was hardly a gullible 'UFO nut' or New Age seeker. Alexander was among the lead investigators of the Skinwalker Ranch. He told Open Minds TV:

"What we learned was that the events (on the Skinwalker Ranch) were certainly real and tangible and definitely occurring. They weren't figments of somebody's imagination or folklore or any of these sorts of things ... but we remain mystified."

Alexander said the 'intelligence' (or whatever it was) on the Skinwalker Ranch seemed to revel in playing mental games with the team of scientists, always one step ahead of anything they tried to do. In one instance, Alexander said the 'intelligence' mysteriously dismantled one of the video monitoring systems on a pole on the outskirts of the cattle pen. Alexander described the incident this way in an Open Minds TV interview:

"The cameras were on top of platforms. They had PVC connectors holding them into the ground ... there was a huge amount of duct tape holding the wires up. At some point, the wires were jerked loose and the cameras stopped recording. All of the duct tape on that post is gone ... not cut ... it is gone. The PVC clamps are gone. The PVC post is pulled back. There is a chunk cut out of the wire, about three feet, which is just missing."

What makes the damage to the camera equipment on the pole even more mysterious is that there were other cameras positioned in the area, including another camera that could see the set-up of the dismantled camera pole. This 'camera watching a camera' recorded nothing happening at the time the deed was done.

Alexander likes to call whatever intelligent force is operating on the Skinwalker Ranch 'a trickster' because it was always one-step ahead of any attempt by the scientists to gather information about it. It also clearly reveled in playing with the minds of whoever was doing any kind of work or experiment – be it ranching or science.

A Night Watchman Falls into a Parallel Universe

So many strange tales relating to the Skinwalker Ranch have been told, and from a variety of sources, that they are almost without limit.

Another highly evocative tale of parallel universe activity at the Skinwalker Ranch comes by way of writer Erick T Rhetts in his book, *Lost on the Skinwalker Ranch*.

Rhetts informs readers that his book tells of true events as experienced by a man who was once employed as a security guard and night watchman on the ranch.

The security guard initially told his story to Rhetts on the condition he remain anonymous. Part of the reason for his need for secrecy is that Bigelow required all workers to sign a non-disclosure agreement about anything that happened on site. They were also forbidden to give press interviews.

However, after leaving his position, the former night watchman left the United States to live as an expatriate in Peru, and subsequently was willing to have his story about the Skinwalker Ranch disclosed.

As do all others who spent considerable time on the lands of the Skinwalker, the former security guard experienced more than his fair share of frightening and bizarre encounters. Again, these included frequent sightings of glowing orbs, ghostly humanoid forms wandering the dark landscape at night, a few UFO sightings, and a number of horrifying encounters with cryptoid beasts, including giant wolves and upright walking 'wolf men'.

It was a close encounter with two such wolf-humanoid creatures that resulted in the most harrowing of experiences - an apparent slip through an inter-dimensional portal and a mind-boggling trip into an alternate world, or perhaps an alternate time scenario.

It all began one night when on patrol. The security guard was making his rounds in the dark, walking carefully among the scrub brush and rocky desert-like landscape of the ranch when he noticed what at first looked like two people walking towards him.

When he shined a powerful light upon the two figures, to his horror he saw not two men but two wolves who were walking upright on two legs, like men.

The two wolf men were tall and skinny, almost emaciated looking. When they spotted the night guard, they began running towards him, and it was immediately apparent that their intent was not for a friendly conversation or to ask for directions. The guard was convinced that these humanoid wolf creatures had evil plans for him, so he turned and ran.

The guard fled in the direction of an ancient abandoned cabin that had existed on the property for many decades. It was basically a rotted out frame of a house with a leaky roof barely in place and nothing more than dirt for a floor. Its windows and doorframes were hollowed out. In fact, the shack would have made an excellent prop as a haunted house for a Hollywood movie as it also had a reputation for strange happenings occurring within it. Many others had reported that odd-looking creatures, ghosts and other figures were seen to enter the cabin, but were never seen to come back out.

Although the security guard was not keen on entering this so-called haunted shack, he felt he had no choice, as the evil-looking werewolves were now hot on his heels.

After a heart-pounding run through the dark, the guard reached the cabin and plunged through the gaping empty doorframe but, just as he stepped inside, he felt as if he slipped, although instead of crashing to the ground he felt himself falling – falling and falling through what felt like dark empty space.

After an indeterminate period of time and confusion, the man sat up on the ground, but clearly could see that he was not inside the hollowed out interior of an abandoned cabin in Utah. Rather, the guard found himself deep inside what appeared to be a vast underground cave.

Stunned, he began to grope his way around, hoping to find a way out. Curiously, he was not ensconced in total darkness. A dim, yellow light emanating from nowhere and everywhere created an ambient illumination, making it possible for him to see just enough to find a narrow passageway. This route seemed obviously human made because it consisted of steps moving upward.

As he made his way through this narrow stairwell, he saw paintings of odd figures and strange symbols on the cave walls. After what seemed like an interminable time climbing through the claustrophobic confines of the cave, he eventually emerged outside, much to his relief. That relief was short lived, however, because he now found himself on the floor of a strange valley surrounded by craggy peaks of stony mountains.

Positioned high in the walls of the canyon the long-lost security guard was stunned to see cave dwellings and milling about in these he could clearly see human figures. Some of the cave dwellings glowed with the soft yellow light of campfires against which could be seen the silhouettes of primitively dressed people.

There were more adventures to come – we recommend Erick T Rhett's book for the full story – but the man eventually found his way back to tell his story. Clearly, however, the tale has all the earmarks of an inter-dimensional slip through a portal that deposited a poor frightened night watchman in another realm, or possible another time.

Sky Portals

Also frequently reported both in the confines of the Skinwalker Ranch and in areas nearby are what appear to be energetic openings in the sky. In his book, *Future Esoteric: The Unseen Realms*, author Brad Olsen writes:

Among the strangest of all the phenomena (on the Skinwalker Ranch) is an occasionally seen large orange portal in the sky, which appears to open into another dimension. At night, blue sky can be seen through it, and black vehicles have been seen entering and leaving the portal ... In addition to teleportation and magnetic anomalies, there also appears to be an intrusion of alternate realities. These parallel universes, or gateways to the higher dimensions are connected to the orange portal. The portal, among a select few others reported worldwide, might be the pathway for cryptozoology creatures coming and going into our third dimensional reality.

A Baffling Mystery Unsolved

Whatever is happening on the Skinwalker Ranch, any attempt to pin it down, identify or quantify it or provide even a simple theory about it has utterly failed. Many of the best minds in the world have visited the site, as have a steady stream of curious trespassers. As Colonel John Alexander, one of the lead investigators with Bigelow's NiDSci team, said:

"Anybody who thinks they understand the true nature of the phenomenon is only fooling themselves."

The Skinwalker phenomenon defeated the best efforts, high tech and otherwise, of the brilliant scientists of NIDSci. As of today, NIDSci is no more. Robert Bigelow pulled the funding and the plug on his pet paranormal investigation team. It disbanded in 2004.

The exception of knowing something for sure might lie with the Native American elders who still live on the land today. For them, the bizarre apparitions and happenings have been a 'natural' part of the landscape for centuries. The Ute and the Navaho may not understand the creatures and skylights of Skinwalker, but they 'know it' and they have learned to live side-by-side with it for centuries on end. They also know this:

It is the best policy to simply never talk about what happens on the Skinwalker Ranch.

In his book, *Skinwalker and Beyond*, author Ryan Burns writes:

Much like other segments of Native American Lore, tribal members protect the secrets behind the issue. However, there is one significant difference, the subject of Skinwalker which is treated with remarkable care. This is not something to be talked about. Many believe even saying the word Skinwalker can awaken the very evil which is being spoken of ... to truly understand the history of the Skinwalker, one must understand the history of the culture which created it.

Even so, with the advent of modern technology and interest in UFOs, space travel and possibly even interstellar travel, time travel and inter-dimensional travel, the possibility that there are key doorways existing in this area will always be a kind of 'catnip' for those people who want to know or who just have to know. They have a need to explore and are ever eager to find those hidden doorways that will lead to other, possibly magical realms.

Thus, interest in the Skinwalker Ranch is not likely to fade away, especially now that the internet and modern media have spread the strange stories of this location around the world. Where there are possibilities and unexplained mysteries, there will always be seekers who will risk it all, and take their chances to discover what they can at the Skinwalker Ranch.

Doorways

Some people call them *Doorways to the Gods*.

They tend to be ancient stone structures, both natural and man-made, and they have long histories of exhibiting strange phenomenon, including the possibility that they are conduits into other realms of reality - or what we may now think of as parallel worlds.

Such doorways can be found all over the world, on every continent, and yes, even in frigid Antarctica. Just looking upon some of the more famous examples of these imposing and massive stone constructions creates a remarkable feeling, even if nothing of note happens and no phenomenon occurs.

Nevertheless, if it is bizarre phenomenon you are interested in, some of the most amazing paranormal events are associated with locations and structures that have come to be known as magical doorways.

The Peruvian Stargate

One of the most amazing examples of a Doorway to the Gods is located in a remote region of Peru, in the Hayu Marca mountain region of southern Peru. The structure is massive. It is a large square construction seemingly carved into an outcropping of a rugged rock formation. Inside the outer frame of the square at the bottom is another smaller indentation that looks just like a doorway.

But it's all solid rock. If this is a doorway, it is no ordinary door. Called by locals Puerta de Hayu Marca (Gateway of the gods/spirits) the structure looks like a door with no way in. It makes one think of those movies where explorers confront a secret doorway or passage and then search for a secret panel or hidden key that will make the door magically slide open.

The discovery of Puerta de Hayu Marca has a remarkable background. The structure was unknown to archaeologists until the 1990s. It was about this time that a local tourist guide, Jose Luis Delgado Mamani, began having strange dreams.

One night, Mamani found himself within an unusually vivid dream in which he was approaching a large stone doorway in a mountainous region. In the dream, there was a pathway paved with pink, polished marble leading up to the door. There were also pink marble statues of exquisite design lining both sides of the path. As he approached closer, Mamani saw that there was a smaller door inset within the larger doorway. Out of the smaller door he could see a shimmering light and when he looked into the light, he could see a vast tunnel that seemed to lead off into infinity.

The dream became a recurring one for Mamani. He told his family about it, and discussed it with friends. Most considered it just a dream, while others said his subconscious mind or higher self was trying to tell him something. Even so, Mamani had no idea what the dream might mean.

About a year later, Mamani was exploring a remote mountainous region he had never traversed before, even though he was an extremely well-travelled guide of the region for mountaineering tourists. The reason for this new deep trek was to find possible new locations for his tourist clients.

As he was walking along Mamani saw something that stopped him dead in his tracks. He was stunned. There ahead of him in a stand of rock thrusting out of the rugged mountain surface was the doorway; that same doorway of his dreams. No, there was no pink marble path lined with statues, or shimmering portal in the centre, but all else was the same. A large square recess carved into the rock along with what was obviously a smaller doorway at the bottom centre.

Mamani told a local Peruvian newspaper:

"When I saw the structure for the first time, I almost passed out. I have dreamed of such a construction repeatedly over the years, but in the dream the path to the door was paved with pink marble and with pink marble statues lining either side of the path. In the dream, I also saw that the smaller door was open and there was a brilliant blue light coming from what looked like a shimmering tunnel. I have commented to my family many times about these dreams, and so when I finally gazed upon the doorway, it was like a revelation from God... How do you make order of such a strange occurrence?"

The discovery of this previously unknown and obviously very ancient monument set off a sensation among archaeologists and historians. Dozens travelled to the location to study the site, leaving all baffled. Few clues remain as to which ancient culture of the region may have constructed the stone doorway but archaeologists speculate it pre-dated the Inca Empire and the structure was probably the work of a forgotten or as yet unknown more ancient culture.

However, historians and anthropologists began looking for clues, knowing that the best place to start was with the local common folk, the descendants of the native races who maintained the lore of the ages by passing down the legends and stories by word of mouth.

The Indians said long ago, far back in the unimaginable mists of time, great heroes used the doorway to pass into an exalted realm where the gods dwelled. They also said that it was possible for these heroes to return from the doorway to spend time back in this world to inspect all the lands of the kingdom.

When the Spanish Conquistadors arrived in the New World in the 1500s to invade Peru and conquer the Inca Empire and loot their gold, it is said that one Inca high priest fled, and his refuge was the Doorway of the Gods. In a fascinating account surviving in rare written manuscripts of the Inca, this priest, who is identified as Aramu Maru, had in his possession a sacred golden disk which was called the Key of the Gods of the Seven Rays.

Aramu Maru used the golden disk to open the Doorway of the Gods, and as the story goes, a shimmering blue light emerged from behind the door. Aramu Maru plunged into the blue light and was never seen again.

Of course, modern scholars write-off this tale as mere mythology and legend of the native Peruvians. However, it is interesting to note that it can still be observed today that the massive stone doorway contains a hand-sized circular depression that looks like a perfect slot for the placement of a circular disk that might serve as a key. Also, people today, both tourists and travellers, report that when they lay their hands on the small door and this circular depression, they can feel a subtle electrical energy. People that are perhaps highly sensitive speak of even more and have reported seeing visions of stars, columns of fire and hearing otherworldly, rhythmic music.

Other modern day witnesses have reported countless strange sightings in the vicinity of the Stargate. These include encounters with extraordinarily tall men and women with fair complexions, blonde hair and arrayed in shimmering robes. UFO buffs point out that these descriptions match a type of alien long known as the Nordics. Often seen emerging from traditional flying saucers or various UFO types, many have come to believe that the Nordics have been on Earth for not thousands of years, but hundreds of thousands of years. Some believe they are a race originally of the earth itself, while others claim they came here from a distant star system. Still others suggest the tall, handsome race of people witnessed in the vicinity of the Stargate are survivors of ancient civilizations, such as Atlantis or Lemuria.

Other phenomenon are observed in the region as well. Golden balls of lights, UFOs of many descriptions and even strange crypto zoological beasts have been seen in the vicinity of the Peruvian doorway.

And then there's this story as told to long-time Minnesota journalist Ken Korczak by a North Dakota man who visited the Andes region of Peru, and spent a night camping just one hundred yards from the door-like structure that locals call the Doorway of the Gods.

In 2012, Duane Sesart, a frequent traveller, hiker and explorer of mountainous regions, made a trek to Peru's most famous ancient monument, Machu Picchu. While there he met up with a group of three other men who said they planned to visit the Hayu Marca Mountains of southern Peru and they invited Sesart along. He eagerly agreed to join the spontaneous adventure with his new friends.

One of the men in the group had heard about the so-called Peruvian Stargate structure, and he wanted to see it for himself. After travelling to the region and hiring a guide, the group of four men set off for the long, steep climb up the primitive mountain trail, finally arriving at the site after two days of arduous hiking ever higher up the stark landscape that he described as both bleak and magnificent.

They arrived at the goal of their destination, the site of the Doorway of the Gods, just as the sun was setting. After a quick initial inspection of the monument, the men scrambled in the dark to set up their camp and prepared for the cold night ahead. In the morning, they planned a closer inspection of the monument and surrounding region.

It was about six hours after dark, in the wee hours of the morning, when strange things began to happen. In Mr Sesart own words:

"We were all exhausted and we pitched our tents. I had my own small tent, and my three friends had set up two tents. One guy slept in one and two in the other. I think I conked out dead asleep within twenty minutes. It was about two in the morning when I was awoken by bright flashing lights I could see through the fabric of my tent. Curious, I poked my head out of my tent and saw one of my friends walking with a bright flashlight - he was walking toward the stone doorway structure.

"I assumed he just couldn't sleep, or perhaps he was just going out to answer nature's call. I thought this explained the lights that had aroused me from sleep, but just as I was about to duck back into my tent, I saw a bright flash in the sky. My first thought was 'lightning' but I quickly knew that couldn't be because the sky was full of blazing stars. It was a crystal clear, calm, cold night.

"Then, to my utter bewilderment, I saw a large globe of light sail across my field of view and veer off to the north, where it quickly disappeared. That was it! I got out of my tent, grabbed my own flashlight, and called to my friend who I had seen walking into the dark with his own flashlight.

"In another minute, I saw a light moving toward me from where my friend had gone, and naturally I assumed that it was him and the light was his flashlight, but I quickly changed my mind. This light was a slightly reddish orb sort of light, about the size of a baseball, and it was kind of bobbing up and down lazily - but definitely moving toward me!"

Mr Sesart said as the light came closer he could see what looked like a human figure underneath it, just faintly illuminated in a dusky reddish glow. He said:

"It was such an eerie moment. I felt frozen with uncertainty. I didn't think this was my friend walking back toward me ... I didn't know what to think."

Presently, the figure of a tall woman walked up to Mr Sesart and stood before him. At some point, the red ball of light above her head had 'somehow just gone away' and now it was only his flashlight that helped him see the woman. And she was extraordinary! He described her this way:

"She was the most beautiful woman I have seen in my life. She must have been at least six-and-a-half feet tall. She was slender and her skin, in the dark, appeared to be very fair, almost a pale white, yet there was a warmth about her. Her hair was red-bronze, short cut, but thick and lustrous. I don't know if it was my imagination, but her eyes seemed maybe half-again as large as that of a normal person. But they were gentle, soft doe eyes with long eyelashes. Her facial features were delicate and she had a slender nose, but full lips. It was hard to tell because I did not want to shine my light directly in her face - that just seemed rude - but her eyes were deep blue, or perhaps a vivid blue."

Strangely, Mr Sesart says he can't for the life of him remember how the woman was dressed, other than to say her entire being exuded a kind of friendly magnificence, or perhaps sense of elegance that was graceful to the point of being actual grace!

At a loss for words and tongue-tied, Sesart finally managed to speak. He said to the woman:

"I didn't know there was anyone else camping out here with us. Are you with a group?"

The woman replied:

"No, I'm on my way … over there," she said, just pointing her finger into the dark.

Again, Sesart felt foolish and at a loss for words, which he finds difficult to explain now. The whole situation seemed eerie, and out of place. There was not only a strangeness about this tall beautiful woman who towered over him, but a kind of atmosphere or 'sphere of strangeness' that engulfed everything around her.

"It wasn't as if I was hypnotized," Sesart said. "It was just an intense feeling of unreality, but that doesn't really describe it either. There was a sort of unique texture of sensation to be in the presence of this … this … who or what was she exactly? I'll be honest, I don't think she was an ordinary human being."

In the next instant, the encounter was over. The woman smiled at Sesart and walked towards where she had pointed into the darkness. It was as if the night had swallowed her up. She was gone.

Sesart went over to the other tents where his friends were sleeping, and was surprised to find all three of them fast asleep. He was certain that he had seen one of his companions walk towards the Stargate with a flashlight just minutes before, but none of the men claimed to have gone anywhere. He told them of his encounter with the tall, mysterious and lovely tall woman.

It was still three hours until sunrise, and the men decided to stay up the rest of the night 'looking at the sky paved with blazing stars' and hoping to see more strange phenomenon, perhaps more balls of light or even a UFO. Duane Sesart hoped to glimpse the graceful, slender lovely 'giant' he had met only too briefly, but she never returned.

It is interesting to note that reports of tall red-headed people are frequent, especially in the mountainous regions of Peru. Consider this report filed with investigative journalist Linda Moulton Howe's website, *Earth Files*:

Location: Pillpinto, Peru. Date 1981. Time: Unknown. In a rugged mountain area travellers and local peasants reported encountering several very tall, almost eight-foot tall beings with copper red hair, crystal blue eyes and very white skin. Local police officers saw a large shiny circular object lift off from the area leaving a burned oval-shaped patch on the ground.

Indeed, the number of reports of tall 'ginger top' people in the Andes regions are too numerous to mention here, and could easily constitute an entire book in and of itself. That four American men on a hike in the Hayu Marca Mountains just happened to encounter a being of the same description near the Peruvian Stargate seems like a lot more than mere coincidence.

Is the Puerta de Hayu Marca, the Gateway to the Gods, a true inter-dimensional portal that leads to a marvellous parallel world? It is certainly a good candidate for that distinction.

Another 'Doorway of the Gods'
In 1956, two young men who were looking for treasure in a remote region of southern Arizona found a lot more than they had bargained for. They found no treasure, but are convinced to this day that they discovered a doorway to a parallel universe, and also a portal that can transport people and objects through time.

Ron Quinn and his brother, Chuck, were residents of the state of Washington. They were avid readers of tales of explorers and treasure hunters. In their readings they uncovered tales of a vast sum of gold hidden or left behind by the Spanish conquistadors as they explored the remote regions of northern Mexico and what is now the southern section of the state of Arizona.

Ron and Chuck Quinn made their way to the tiny village of Arivaca, Arizona, where about seventy people lived in relative isolation from the world. Over the next several days, the brothers trekked out into the rugged desert landscape and discovered they had their work cut out for them. This was an area of endless desert and rock landscape for miles in all directions, which was fine with the Quinn brothers. Treasure or no treasure, exploring the rugged landscape was an adventure in itself.

On their second night of camping, Ron and Chuck were enjoying the sunset of the desert. In the distance was the outline of the craggy peaks of the Tumacacori Mountains. The stars were just coming out when the men saw something extraordinary - two luminous objects of large balls of blue-green light appeared to be 'coming in for a landing' behind the mountains. Ron said the objects definitely were not conventional aircraft or military flares. The desert evening was silent and they heard no sound of manmade aircraft. They noted the time was five minutes after eight o'clock.

The next evening at exactly the same time the mysterious green orbs made a second appearance. Neither Ron nor Chuck were UFO buffs at the time, although they were well aware of the phenomenon. After all, this was the 1950s, sometimes called the 'flying saucer fifties' because of the explosion of sightings across the U.S. in that decade.

A few days later when visiting the village of Arivaca, Ron and Chuck met a local rancher and long-time resident of the area. They told him about the strange light they had seen but the rancher seemed not at all surprised. In fact, he told the brothers that these glowing orbs had become almost a common sight, beginning in 1939.

The Archway Portal

One of the more intriguing natural formations the Quinn brothers discovered in the desert was a stone archway carved out by Mother Nature from one of the many large rocky outcroppings that thrust up from the desert floor. Such formations, although not numerous, are well-known features of the American southwest desert-mountain landscape.

On one of their trips from town back to the area of exploration they had mapped out, the brothers came across a local Native American man whose pick-up truck had stalled on the side of an old dirt road. They offered him assistance and a ride into town. On the way, the brothers mentioned the natural stone archway they had discovered in the desert. The Indian asked them, rather cryptically:

"You didn't try to walk through it, did you?"

The men said they had not. The Indian said:

"Well, you should know my people have considered that archway a doorway into another world; it's a story that has been handed down through the centuries. The white people around here call it Indian superstition, but many have learned different."

The Indian man told them of a common tale among his people about a young brave who was seen to walk under the archway, but never emerge again on the other side. His companions searched for him in vain, and returned to their Indian village. They consulted with their tribal medicine man who told them that the archway was an 'ancient portal' and demonstrated the fact by tossing a live rabbit into the archway, only to see the critter vanish just like the Indian brave who was never seen again - or was he?

In the 1940s a local rancher named Louie and a fellow ranch worker were searching for a lost bull in the vicinity of the arch, when they happened upon a set of human remains. The body was mostly bones with some dried skin attached, but its head and one leg were missing. What was even stranger however was that still attached in tatters to the remains were what appeared to be old Indian buckskins, the kind not worn by Natives for more than one hundred years. They also noticed that next to the body was a bow-and-arrow made of wood, and the wood was in an advanced state of disintegration. It seemed that the ancient weapon was obviously of the kind used in previous centuries by members of local Indian tribes.

Certainly, the body had been on the site decomposing for a long time, but not the two centuries or so that would have placed this specimen in its proper time period. This body appeared to be the remains of an Indian man who had been decomposing in the desert for a few decades. Could it have been the young brave who had vanished through the archway so long ago, only to return a century or two later in the wrong time?

Even among the local Indians of today, there is debate as to whether the stone archway is of special significance or not. Still, some people sometimes go there to 'test the doorway' by throwing in rocks or by daring to jump back and forth through the structure. Most often, nothing happens, but it seems that when conditions are just right, strange events can occur. Many people pass under the archway, for example, and although they do not vanish into another dimension, they come away confused and disoriented. Most of the time a 'missing-time' experience is reported - that is, just walking through the doorway causes a person to feel they have been gone for hours or days at a time, when they clearly went nowhere and only appeared to pass through the arch without incident.

Tom and Chuck Quinn found the lore surrounding the stone archway interesting, but also immediately filed it away as a colourful local legend. Nevertheless, over the years Tom, his brother and their friends have revisited the site on numerous occasions. Tom relocated from Washington to live in Tucson, and the harsh and barren landscape he enjoyed exploring as a young man in his twenties has remained a favourite getaway for exploration, camping and fun ever since.

One of the aspects of the site, which Tom considers significant, are the large number of geodes that can be found near the archway. Geodes are a special kind of rock of sedimentary origin, and are formed by chemical precipitation. They are essentially hollow, and vaguely spheroid in shape. Many rock hounds and collectors value them because when you crack them open they often contain beautiful bunches of various crystal substances, such as quartz, agates, jasper or chalcedony. A geode can be mostly hollow but lined on the inside with a colourful crust of crystal, or be filled with crystal nodules.

What is strange, however, is that Tom has noticed that piles of geodes seem to appear or vanish as if by whim from the archway vicinity. On one visit to the site, Tom was surprised to find a huge geode about ten feet in diameter near the arch in a location he had been to dozens of times before, and yet he had never before noticed this massive, remarkable specimen. He is convinced that geodes appear and disappear from the site at random.

On another occasion, Tom had a more remarkable experience. In October of 1973 Tom was hiking again in the desert valley, and had spent the better part of an afternoon climbing up a canyon wall. The climb was arduous and precarious, involving slipping and sliding among boulders and various areas of loose rock and incredibly rugged terrain. When he sat down to rest, Tom was in for a surprise. He told the *Tucson Weekly*:

"It's a long weary climb, so I paused for a breather halfway up. I sat on the slope facing north. To my left (west) the steep hill followed the canyon perhaps a mile, but something was definitely wrong. Below to my left was a canyon where none had existed. Curious, I made my way down, entering it from the eastside, so I thought.

"I soon discovered I was in the same canyon that led toward the hill I had just scaled. I was more than two hundred and fifty yards back down the canyon on a different slope and now I was facing south - I had mysteriously been transported to the new location. Thinking I was looking west, I was really looking east seeing the canyon I had just hiked."

Many Strange Sightings

Tom Quinn is not the only person to witness strange events in this remote corner of the world. Even though very few people trek into this bleak corner of the American southwest, those few who visit here have a remarkable tendency to see something unexplainable.

For example, in the 1960s another local rancher said he once spotted a column of walking men about two hundred yards below him from his vantage point on a hill within the valley. Taking out a pair of binoculars, the rancher was stunned to see about thirty men dressed like 16th Century Spanish soldiers. The men, walking single file and carrying lances wore helmets and long white shirts with leather vests or belted with leather. They had two horses with them which were packed with supplies and indeed the out of place group looked for all the world like a patrol of Spanish conquistadores. Tom thought perhaps someone was filming a scene in a Hollywood movie, but quickly abandoned that idea when he noticed no cameras or film crew in the vicinity.

Another local resident tells of the time when he was crossing the region on a cloudy day when a storm had been threatening to overtake the region. When it started to rain, the man took shelter under a rock shelf very near the stone archway. As the rain grew worse and a fierce thunderstorm broke out, the man was surprised to see that, through the arch, he could see a blue sky hovering over a sunny day! It was as if the arch was a doorway which was acting as a window showing the same valley, but on another more pleasant day, or perhaps a time from any other given century.

There have been numerous reports made by hikers and campers of hearing the thunder of hoof beats, as if large herds of wild horses are nearby and coming towards them, only to discover an empty, silent valley when climbing higher for a look around. Large herds of horses were common within this region in centuries past, but do not exist there today. It almost seems as if the flow of time is precarious and unstable in this vicinity, and there is something about the landscape that causes different eras of time to waft and flow into each other at random. Some speculate there are powerful, but natural sources of electromagnetic activity deep beneath the surface of this area, which could somehow be 'bending time'.

Of course, that would not explain the numerous UFO sightings. Just as Tom and Chuck Quinn observed large green UFO orbs on two successive nights in the region, dozens of others have reported UFO sightings of numerous kinds; in fact, some UFO buffs now venture out to the site solely for the purpose of UFO spotting or hoping to get a glimpse of something otherworldly.

The vast, lonely deserts of the southwest remain sparsely populated to this day, and many maintain it remains a favourite landing site for extraterrestrial visitors from points near and far in the vast universe of galaxies where endless numbers of spacefaring civilizations have had plenty of time to develop capability to travel among the stars.

Some are worried, however, that this special region is in danger. Plans are going forward by the Tucson Electric Power company to build a 345,000-watt high voltage transmission line that will travel between Tucson and Nogales. The structure of power lines will create a considerable electromagnetic signature in the valley very near where the archway stands. What will happen when this modern form of manmade electromagnetic 'pollution' interacts with the endemically strange powers of the region? It has many people worried. The effect could be catastrophic, or might simply drive away the amazing paranormal phenomenon by interfering with the natural balance of forces in this region.

If the project goes forward, only time will tell. A natural portal between other worlds and other times may be closed for good or may be thrust open even further.

Fata Morgana

Newspapers and internet news sources went crazy in October of 2015 after literally tens of thousands of Chinese citizens witnessed an amazing event in their sky. They saw the image of a magnificent city floating in the clouds thousands of feet overhead.

Yes, there were skyscrapers in the sky!

Pictures were taken and video footage was shot, and soon the dramatic images were circling the globe. If the whole event had not been captured by dozens of cameras and a few video clips, the story might soon have disappeared as mere hearsay or simply a strange rumour among the Chinese people.

But the images don't lie. Anyone can clearly see this remarkable vision of high-rises on high - although that has not stopped many sceptics from calling it all but a lie.

Material science literalists at first insisted in dismissing the photographs and videos as hoaxes and trick photography. However, this did not square with the tens of thousands of people who were in Foshan, China, on that day in October.

Then, a few days later, over the skies of Jiangxi, China, yet another floating metropolis appeared, riding aloft in the atmosphere as tens of thousands looked on in wonder.

Conspiracy theorists and fans of all things paranormal chimed in with all kinds of explanations, perhaps the most common of which was that the floating city was a glimpse into a parallel universe.

Somehow, some said, a rift had opened up in the veil between universes which allowed thousands of people a glimpse of a city that is not of this world. Maybe it was a brief, freaky window opening into one of the billions (or infinite number) of alternate worlds which even most scientists agree today makes up the composition of our 'multiverse'.

But the sceptics were not done. After learning that the amazing images in the sky could not have been a hoax or trick photography, they reformed their ranks and pounced. This time they abandoned their notions of fake CGI-enhanced videography, and eagerly offered an 'obvious' scientific explanation for the bizarre sightings of cities in flight. They said the phenomenon was actually well known and had been observed for centuries. They said it was a classic 'Fata Morgana'.

In short, a Fata Morgana is a kind of naturally occurring optical illusion created when conditions are just right for the light rays in the atmosphere to bend, sending or projecting images of faraway places to another location, making it look incredibly real and solid to observers on the ground.

Naturally, when mainstream scientists trotted out the Fata Morgana explanation based on natural atmospheric tricks, the vast majority of the general public considered the whole event 'case closed'. This seemed such a valid and suitable explanation and it satisfied everyone's 'comfort zone'. Ahhh ... hard science comes to the rescue again!

The problem, however, is that the explanation offered by sceptical science makes no sense. Just take a look at what the first sentence the *Wikipedia* online encyclopedia has to say about the Fata Morgana:

*A Fata Morgana is an unusual and complex form of **superior mirage** that is seen in a narrow band right above the horizon.*

Notice that it says a Fata Morgana mirage is seen as a *narrow band right above the horizon.* That hardly describes the floating cities above two different urban centres of China. These apparitions were high in the air at greater than 45 degrees beyond the horizon. That would make it impossible for the laws of light and refraction principles to produce these stunning images.

But the sceptics were right about one thing. The phenomenon that has come to be known as the Fata Morgana has been known for centuries. The people of China were by far not the first to be stunned by the image of a fantastic city floating up in the middle of nowhere. Before we discuss the possible parallel universe connection of what has come to be known as the Fata Morgana, let's take a look at some of the most famous sightings throughout history.

In 1508 an Italian doctor, scholar and humanist by the name of Antonio de Ferrarsis published a book titled, *De Situ Japigiae,* which tells of his observation in a region of Italy now called Apulia, which is in the 'heel' area on the boot of the Italian peninsula.

In it he describes a frequent phenomenon in and around the Swamps of Taranto where strange sightings were often reported by locals. In his book, he writes:

... certain apparitions are seen, which are called Mutationes or Mutata. The common people tell tales of I don't know what, vampires or witches or, as they say in Naples, janare [fairies], or as the Greeks say, nereids ...

And sometimes you will see cities and castles and towers, and sheep and different coloured cattle and images or spectres of other things, where there is no city, no sheep, not even a thorn bush. I myself have sometimes had the pleasure of seeing these plays, this lusus naturae.

Another brilliant man of Italy who lived just shortly after de Farrarsis was the Dominican friar Tommaso Fazello, today often called the Father of Sicilian History. He also made note of the occasional apparitions of strange floating cities in a similar region of southern Italy, but especially over the sea between Calabria and the island of Sicily. He writes:

In this same sea is seen yet another wonderful thing, which is that when the storm ceases, and the air becomes still, at dawn, changing images of animals, and of men, are seen in the air, some of which are quite motionless, some run through the air, some fight among themselves, and last even when the Sun gains strength, in whose heat all disappear.

In 1658, Athanasius Kircher, a Jesuit Priest and scholar, wrote in his book, *Ars Magna et Umbrae*, about a fabulous instance of Fata Morgana he observed on 14th August 1643, on the Straight of Messina:

On the fifteenth of August, 1643, as I stood at my window, I was surprised with a most wonderful, delectable vision. The sea that washes the Sicilian shore swelled up ... while the waters near our Calabrian coast grew quite smooth, and in an instant appeared as one clear polished mirror, reclining against the aforesaid ridge.

On this glass was depicted, in chiaroscuro, a string of several thousands of pilasters, all equal in altitude, distance, and degree of light and shade. In a moment they lost half their height, and bent into arcades, like Roman aqueducts. A long cornice was next formed on the top, and above it rose castles innumerable, all perfectly alike. These soon split into towers, which were shortly after lost in colonnades, then windows, and at last ended in pines, cypresses, and other trees, even and similar. This is the Fata Morgana, which, for twenty-six years, I had thought a mere fable.

There are many more examples from medieval accounts to the post-medieval years, and right up into modern times. Let us jump forward in time to the early 1800s when Irish writer Anne Plumptre gave this account in her book, *Narrative of a Residence in Ireland during the summer of 1814*:

It was in summer evenings, when the clouds appeared remarkably electric; appearances then exactly resembling castles, ruins, tall spires, groves of trees, rocks, and other terrestrial objects seemed to sail rapidly along the surface of the sea from the east to the west, remaining for a length of time sufficient to give assurance that such appearances actually existed, that they were not the mere effect of strong imagination; —at sunset they wholly disappeared. ... In 1748 a book was published by a gentleman residing near the Giants' Causeway, in which a curious detail is given of an enchanted island seen annually floating along the coast of Antrim, of which it is said a sod thrown on it from the terra-firma would fix it for ever.

Crocker Land

One of the most famous examples credited to the Fata Morgana phenomenon is the long and complicated saga of Crocker Land and/or Crocker Mountain.

Some dismiss the whole 'Crocker Affair' as an out-and-out hoax, while others insist it was an incredible real and persistent Fata Morgana mirage which just happened to be so powerful that it convinced some of the world's greatest Arctic explorers that a vast mountainous island existed in the Arctic Ocean northwest of Canada.

Others maintain that the Crocker Land 'illusion' was just too powerful to have been a mere mirage, and a better explanation is that its legend represents an actual window into a parallel universe. But before there was a 'Crocker Land' there was 'Crocker Mountain'. Let's explain, starting at the beginning:

The year was 1818 and English explorer Sir John Ross was leading a voyage seeking to discover the long sought after Northwest Passage.

After reaching Lancaster Sound, the Northwest Passage was ahead but, unfortunately, Ross sighted a massive landmass complete with high-peaked mountains covered with snowy caps looming further to the northwest. The idea that this was a momentary mirage or a shimmering Fata Morgana seemed an unlikely explanation. That is because the vision persisted for so long, some say days and perhaps weeks and was simply not consistent with weather patterns and atmospheric anomalies. Whatever the case, Ross determined that any further sailing in that direction would be impossible because, to him, the Northwest Passage was blocked by a massive island. He returned to England, his mission an expensive disappointment.

If this long and arduous failure was not enough, Ross compounded the negative consequences of the situation in a most unexpected way in that he named the landmass he was sure he had seen and called it the 'Crocker Mountains' after the First Secretary of the British Admiralty, John Wilson Crocker.

Not only had Ross scuttled an expensively financed expedition, but he had named a 'phoney island' after the great naval officer John Crocker, to the Admiral's great embarrassment. The result was that Ross was never again able to gain government support for expeditions, and he became the object of ridicule in some circles - all because he and his entire crew were convinced that they had seen a massive landmass in the Arctic Ocean, where we now know that no such landmass exists.

But did Ross have the unfortunate experience of glimpsing a rift in our universe whereby he caught a glimpse into a very real, very solid parallel world? Perhaps. However, there is more to the story.

Interestingly, the story of the stubborn and persistent 'Crocker' landmass would strike again almost a century later.

An Incredible Coincidence

Almost ninety years after the Sir John Ross expedition, another famous Arctic explorer, the American Robert Peary, was sailing in the vicinity of Cape Thomas Hubbard when he saw in the distance a landmass which fitted the exact description as described by Ross nearly a hundred years previously. It was a huge island or small continent graced with high snowy peaks and a range of rolling mountains.

Peary was not aware of the Ross story but, amazingly, also named the landmass he was convinced was real 'Crocker Land'. He named it thus in honour of an entirely different man, this time the San Francisco banker George Crocker, who was a major financial supporter of Peary's expedition. Like Ross, Peary concluded that a suitable Northwest Passage could not exist for the obvious reason that the way was blocked by a small continent, or a very large island.

However, some claim that Peary knew full well that there was no real 'Crocker Land'. Some historians say that he fabricated the entire story to secure further funding from the wealthy banker George Crocker. Another explorer, Frederick Cook, had traversed the same region where the alleged coordinates of Crocker Land should exist, but Cook found only empty sea and ice in the region.

Since Cook and Peary were bitter rivals, it was easy for him to suggest that Peary was a fraud and a hoaxer. Many accepted this as a good way to explain away the return of the mysterious Crocker Land after Ross had been thwarted by it in 1816. However, backers of Peary and many others did not believe Cook. They knew Peary to be a man of integrity, not a liar, and that he could not possibly have believed he could get away with such an audacious hoax. Peary's convictions were based on what he saw with his own eyes. This was so convincing, he was able to obtain financing for another expedition to not only locate Crocker Land, but explore its environment. Many others believed the same.

Newspapers of the day were excited about the prospect of an expedition to Crocker Land, with one article quoting the explorer Donald Baxter MacMillan as saying:

"(Crocker Land's) boundaries and extent can only be guessed at, but I am certain that strange animals will be found there, and I hope to discover a new race of men."

The story of the actual expedition to Crocker Land is enormously complicated and is also a tale of incredible suffering and bravery by the intrepid explorers who risked everything to find a landmass that never existed in actuality (at least not in our universe). There are tales of frostbite, death, ships stranded in ice, starvation and even a grisly murder.

Needless to say, when all was said and done, the expedition ended in failure. Again, it turns out that there is no Crocker Land – not in our reality. The result was another considerable financial loss that ruined the lives and reputations of more than a few enterprising investors and fearless explorers.

Historians have long relegated the complex story of the Crocker Land expedition to a case of either outright deceit and deception (that Crocker Land had been a hoax from the beginning perpetrated by Robert Peary) or that it had been an incredible case of an extremely persistent mirage, a Fata Morgana, that had somehow fooled some of the world's most experienced seamen and explorers.

Paranormal speculators today say that Crocker Land must have been more than a simple mirage. It almost certainly could not have been a hoax, the latter which would have fooled dozens of speculators, inventors and other explorers who had been to the region, and who could have dispelled any false claims with ease.

Rather, some theorists say, that Crocker Land is real, except that it does not exist in our dimension, but rather in a very real landmass of an alternate dimension or parallel universe. When conditions are right, it somehow becomes visible from time to time in this remote region.

Supporters of the parallel universe theory point out that even the native Inuit peoples of the region have long spoken of the mysterious landmass in their legends and lore. The Inuit call the sighting of the ghostly island a 'Poo-jo' which some scholars have translated as 'mist', but others suggest that the term has deeper meaning than just 'mist' and have translated it to mean 'illusion'.

The native Arctic dwellers may agree that this ghostly island is something that might not be of this world, but this was not synonymous for them to define it as 'unreal'. To the Inuit, this island is an 'abode of the god' or where spirits dwell, and is a very real place indeed. They even have tales of Inuit tribesman who were able to visit the ghostly, elusive continent and return with amazing stories of adventure, weird creatures and strange peoples.

While ancient peoples define what they have clearly seen for centuries in their terms, those today armed with the knowledge of modern science and quantum mechanics can apply their own meanings - and that may mean parallel worlds or parallel universes.

Can We See Across the Divide?

Scientists seem firm in their conviction that parallel worlds exist, but they also maintain that our ability to see into them is impossible - well, almost. Famous physicist, Machio Kaku, certainly admits that parallel universes hold an extreme possibility of being real because they are predicted by the mathematics of quantum mechanics.

Nevertheless he says that to reach one of these alternate worlds, an enormous amount of energy would be required (this is called the Plank Energy).

Scientists have been telling us for centuries what is impossible, until it becomes possible. It used to be a canon of science that the earth was flat and the sun revolved around the earth. That proved to be wrong. Scientists said that an ancient fish called a Coelacanth had been extinct for more than sixty-six million years until a couple of fisherman hauled one in from their nets off the coast of South Africa in 1938.

One of the world's greatest scientists of his day, Lord Kelvin, said in 1899:

"Radio has no future. Heavier than air flying machines are impossible. X-rays will prove to be a hoax."

Also, science has never stopped ordinary people from seeing and experiencing the impossible. For example, perhaps since the beginning of time people have reported 'falling into' other worlds temporarily, and returning with strange tales. A classic example are those who report trips into the 'Land of Faerie'. Others claim that glimpses of ghosts and other apparitions are not so much visitations by the dead, but a peek into a universe parallel to our own.

And then there are the Fata Morgana, which have been reported for centuries. Again, it's just too easy to jump on the most practical scientific explanation - that a Fata Morgana is a mirage caused by light bending in unusual ways when - temperature and other conditions are just right.

It is these kind of explanations, although certainly legitimate in some cases, which are letting scientists get away with murder, so to speak. The example of Sir John Ross and Admiral Robert Peary being baffled by the impossibly persistent vision of a giant continent would seem to fly in the face of the Fata Morgana explanation. One of the primary reasons for that is in order to create such an illusion there would have to be a real set of mountains somewhere within a few miles (or perhaps a few hundred miles at the most) to give form to the image in the first place! Remember, a Fata Morgana is created when the image of *something real* is projected to another place and is seen where it should not be. If Ross and Peary both saw a Fata Morgana projection of Crocker Land, then where was the original? The answer is, there is no candidate for this proposition.

We can't prove that these Arctic explorers were seeing into a parallel world, but there is currently no acceptable scientific way to explain it away either, no matter what the sceptics might tell you.

The same still stands for the images of floating cities witnessed by thousands over two cities in China, and the multitude of other visions of not only cities, but marching armies, moving animals, herds of Centaurs (as reported by Violet Tweedale in her book, *Ghosts I Have Seen*) not to mention UFOs, cryptozoological beasts such as Bigfoot and a whole raft of other strange monsters, including perhaps one of the most famous of all - the Loch Ness monster. A glimpse into a parallel universe where large amphibious dinosaurs still exists might yet be the best explanation for Nessie.

So it is likely that scientists and sceptics will continue to insist that catching a glimpse into a parallel universe is impossible - and yet people everywhere will continue to do just that.

Conclusion

It has been more than sixty years since physicist Hugh Everett III first proposed his Many Worlds Interpretation and scientists have been arguing about it ever since. This is because the implications of 'Many Worlds' is just so mind boggling. Not only is our universe or dimension of reality not the only dimension of reality, but it's just one of an infinite number of other parallel worlds and, yes, all of those 'other' worlds are tucked in right next to ours. Perhaps only the thinnest of veils separates us from all the rest.

Many scientists still say the Many Worlds theory carries too much metaphysical baggage. On the other hand, a survey conducted of a cross section of the physicists around the world just ten years ago showed that 68% of them now agree that the Many Worlds Interpretation is either 'true' or 'mostly true'.

So now it seems that even the most mainstream of scientists agree that we live in a universe teeming with parallel worlds. Perhaps fewer are on board with time travel but, as we have seen, the two always seem to go together, or at least have something to do with one another.

In the final analysis, it's not science that is going to convince the world that parallel universes are real and that time travel is possible. Sooner or later, the truth is going to be undeniable. When enough people have real-life encounters with a time-slip or a slip into another dimension, sooner or later a critical mass will develop until a consensus is reached.

That consensus may very well be:

Of course we live in a Multiverse; yes we can have contact and interaction with all of those other parallel worlds, and that also means we can travel in time. We know it is true and possible because it happens every day.

***

I hope you enjoyed this sixth book in my Time Travel / Parallel Worlds/ Lucid Dreaming series. If so, please leave a positive review. It will be much appreciated.

If you are interested in reading my other Time Travel, Angel and Heaven books, please visit my Author Central Page for further information:

USA: http://www.amazon.com/Richard-Bullivant/e/B006U1CYRA/

UK: http://www.amazon.co.uk/Richard-Bullivant/e/B006U1CYRA/

www.ingramcontent.com/pod-product-compliance
Lightning Source LLC
Chambersburg PA
CBHW030703190526
45164CB00004B/373